DOCUMENTARY MAKING FOR DIGITAL HUMANISTS

i

Documentary Making for Digital Humanists

Darren R. Reid and Brett Sanders

https://www.openbookpublishers.com

© 2021 Darren R. Reid and Brett Sanders

This work is licensed under a Attribution-NonCommercial 4.0 International (CC BY-NC 4.0). This license allows you to share, copy, distribute and transmit the text; to adapt the text for non-commercial purposes of the text providing attribution is made to the authors (but not in any way that suggests that they endorse you or your use of the work). Attribution should include the following information:

Darren R. Reid and Brett Sanders, *Documentary Making for Digital Humanists.* Cambridge, UK: Open Book Publishers, 2021. https://doi.org/10.11647/OBP.0255

Copyright and permissions for the reuse of many of the images included in this publication differ from the above. This information is provided in the captions and in the list of illustrations.

In order to access detailed and updated information on the license, please visit https://doi.org/10.11647/OBP.0255#copyright. Further details about CC BY-NC licenses are available at http://creativecommons.org/licenses/by-nc/4.0/

All external links were active at the time of publication unless otherwise stated and have been archived via the Internet Archive Wayback Machine at https://archive.org/web

Digital material and resources associated with this volume are available at https://doi.org/10.11647/OBP.0255#resources

Every effort has been made to identify and contact copyright holders and any omission or error will be corrected if notification is made to the publisher.

Open Field Guides Series, vol. 2 | ISSN: 2514-2496 (Print); 2514 250X (Online)

ISBN Paperback: 9781800641945
ISBN Hardback: 9781800641952
ISBN Digital (PDF): 9781800641969
ISBN Digital ebook (epub): 9781800641976
ISBN Digital ebook (mobi): 9781800641983
ISBN Digital (XML): 9781800641990
DOI: 10.11647/OBP.0255

Cover image: Photo by Nathan Dumlao on Unsplash at https://unsplash.com/photos/McztPB7Uqx8. Cover design by Anna Gatti.

Contents

	Introduction	1
1.	**The Humanist Auteur**	5
	The Digital Wave (and the Power It Gives Us)	7
	Film as Scholarly Tool	11
	The Filmmaker-Scholar	14
	The Filmmaker-Scholar as Auteur	19
	Looking for Charlie	23
2.	**Learning to Love the Camera**	25
3.	**The Production Process**	29
	Pre-Production	30
	Production	33
	Post-Production	37
4.	**Concept and Planning**	41
	Schema One — Essay Films	42
	Schema Two — Discussion/Interview Films	43
	Schema Three — Full-Production Films	44
	Schema Four — Subjective Explorations	44
	Achievability	45
	Case Study — Signals	48
	Planning	51
5.	**Collaboration**	55
6.	**Precedent**	63
7.	**Choosing Your Filmmaking Equipment**	75
	Smartphone Kit ($100–1,000)	76
	DSLR Kit ($300–5,000)	77
8.	**Core Methods**	81
	Stabilise your Camera	81
	Focus your Camera on your Subject	82

	Compose your Shots	83
	Plan to Capture Contextual Footage	86
	Take Control of your Camera's Settings	91
9.	**Settings, Lenses, Focus, and Exposure**	**93**
	Camera Settings	94
	Lenses	95
	Stylised Focus	98
	Exposure	99
	Summary	102
10.	**Composing a Shot — Tips and Techniques**	**103**
	Head Room	104
	Looking Room	107
	The 30° and 180° Rules	107
	Fundamentals	110
11.	**Shots and Compositions Considered**	**111**
12.	**The Visual Language of Cinema**	**119**
	Frame Rate	120
	Vulnerability, Strength, and Significance through Camera Angles	120
	Wide Shots, Close-Ups, Mid-Shots	123
	Aspect Ratios	124
13.	**Interviews**	**127**
	Oral History and Interviewing	128
	Designing an Interview	130
	Formulating Interview Questions	132
	The Phrasing of Questions	133
	The Role of the Interviewer	133
	The Interviewer/Subject Relationship	135
	The Ethics of Interviewing	137
	The Interview Process	138
	Participant Information Sheet Template	140
	Informed Consent Form Template	142
14.	**Recording Audio and Creating Soundscapes**	**145**
	Recording Sound on Site	146
	Rough and Ready	146
	Lavaliere Microphones	147
	Run and Gun	147
	Clipping	148

	On-Site Tips	148
	Engineering Ambience	150
	Voice-Overs and Commentary	152
15.	**Light**	**155**
	Core Rules	155
	Hard Light and Soft Light	156
	Lights and Lighting	160
	Lighting Quick-Reference Guide	162
16.	**Camera Movement**	**165**
	Going Handheld	165
	Handheld Tracking	168
	Camera Pans and Tilts	169
	Dolly Shot	170
	Tripod Dolly	173
17.	**The Two-Page Film School**	**175**
18.	**Post-Mortem: Making a Short Documentary about the 2016 Presidential Election**	**179**
	Pre-Production	182
	Production	184
	Post-Production	187
	Aftermath	189
19.	**Post-Production Workflow**	**193**
	Review your Footage	196
	(Re)Consider your Audience's Relationship to the Film	196
	Plan a Working-Structure for your Film	197
	Begin Creating a Rough Cut	198
	Step Back	199
	Critically Review and Reassess	199
	Post-Mortem	200
	Refine Your Rough Cut	200
	Step Back, Review Your Fine Cut and Reassess	201
	Reflections	201
20.	**The Three-Act Structure**	**203**
21.	**The Protagonist**	**209**
	Harmon's Story Embryo	211
	Casting the Audience as the Protagonist	213
	The On-Screen Protagonist — The Journey	217

22.	**Assembly**	**221**
	Editing	221
	Colour-Grading	227
	Sound-Tracking	229
23.	**Editing Workflow in Adobe Premiere Pro**	**233**
	Step One: How to Start a New Project	234
	Step Two: Get to Know the Premiere Workspace	235
	Step Three: Import Video and Audio Clips into your Project	236
	Step Four: Move Clips into your Timeline	238
	Step Five: Shorten a Clip	240
	Step Six: Moving Clips Around the Timeline	242
	Step Seven: Cutting Between Clips	242
	Step Eight: Remove Unwanted Sound-tracks	245
	Step Nine: Add a New Sound-track	246
	Step Ten: Add On-Screen Text	248
	Step Eleven: Saving Your Project	250
	Step Twelve: Exporting Your Project	251
	Step Thirteen (Optional): Colour-Grading Your Project	253
24.	**Distribution and Dissemination**	**255**
	Theatrical Release	256
	Digital Streaming	260
	Freely Accessible Digital Streaming	262

Bibliography — **267**

Illustrations — **277**

Introduction

Documentary films have always been powerful. Robert J. Flaherty's *Nanook of the North* (1922) shaped how entire generations conceptualised Inuit peoples. Almost a hundred years later, Michael Moore's *Fahrenheit 9/11* (2004), for all its faults, was a clarion call for dissent that was heard around the world. Throughout most of cinema's history, however, the documentary has been the purview of a privileged few. To create even the most rudimentary film required access to a vast array of expensive equipment, specialist skills, and traditional distribution models — not to mention the significant financial resources required to fund the purchase of, among other things, expensive film stocks. As the twentieth century gave way to the twenty-first, however, a digital shift has brought audiences online whilst simultaneously providing creators with access to a range of new, easy-to-use, and affordable filmmaking tools. It is now entirely possible, even desirable, for humanists and other academics to utilise the documentary medium for their scholarly purposes. New audiences can be reached, and opportunities to conduct and present one's research using the grammar of cinema and the moving image, previously inaccessible, are now widely available.

For humanist scholars, the potential of this technological development to challenge the traditional format of the thesis, and to engage in new types of research and intellectual dissemination, is staggering. Eye-witness testimony, unfolding events, and oral histories can be recorded, contrasted, compared, and shared. Such materials can be fused with academic commentary, archival footage, and other audio-visual texts to create works that are far larger than the sum of their parts. Singular events can be explored from a multitude of perspectives, underlining, should one choose to do so, the subjectivity of truth. With the correct skills, humanist scholars can now create works and disseminate their findings in new and exciting ways. All that remains is for them to develop the wide range of necessary skills which will allow them to take advantage of these new opportunities.

Documentary Making for Digital Humanists has been created to empower academics, scholars, students, journalists, and other thinkers with the tools necessary to turn their research into intellectually rich films. This book aims to remove the skill deficit that is likely to be faced by so many. It has been designed to take humanist thinkers with little to no filmmaking experience and teach them, in a logical and easy-to-follow manner, how they can create documentary-style pieces of their own. It will take readers through the three key stages required to turn their research into a film: pre-production (chapters 1–7), production (chapters 8–17), and post-production (chapters 18–24). In each section, readers will learn the key ideas, techniques, and methodologies necessary to create scholarly films. In some places this will mean engaging in theoretical discussions about the nature of the field, storytelling, and collaboration; in others it will mean learning practical skills, from setting up cameras to shot composition and the recording of audio. Whether practical or theoretical, this book aims to make the journey from scholar to filmmaker as intuitive and accessible as possible.

To that end, this book combines its text with a ten-part video course. This video course can be accessed from within the pages of this book, by clicking on the play icon of the embedded video in the online edition or scanning the QR codes in the print/PDF edition.

Readers can choose how they wish utilise this text. They can, for example, watch the video course first; or read the text in its entirety; or work through both in tandem. We recommend readers choose one of the following three ways of engaging with this work:

- *Documentary-making course*: this book and its integrated video course has been designed to act as a complete learning experience. By reading the book and engaging with each video lesson (and carrying out the assignments contained therein), you will be walked through the filmmaking process in discrete stages. Assignments issued as a part of the integrated video series will help you to develop practical experience alongside a growing portfolio of filmed material. This approach to the text requires readers to engage with each element of this book (and video course) in order, completing assigned tasks and practising associated skills and techniques. This design would also, with context-appropriate adjustments, function well if integrated into traditional learning environments with lessons and discussions which can be easily mapped against most ten-to-fifteen-week semesters.

- *Quick immersion, long-term development*: to achieve quick immersion into the world of scholarly filmmaking we recommend first watching the video course and then reading the main text in this book. This method of engagement will first provide aspiring documentary-makers with an overview of the filmmaking process before providing them with an opportunity to build upon this core knowledge through more in-depth discussions. For those readers looking to begin experimenting with the medium as quickly as possible, this approach is likely to be the most suitable, with the video course providing necessary core skills, whilst the main text provides opportunities to develop those skills in-depth.

- *Reference guide*: documentary-makers in the field must balance a wide range of responsibilities, skills, and methodologies. For those ready to enter the field, this work can serve as an important point of reference, providing timely, practical insight as well as the workflows necessary to achieve specific day-to-day production tasks, when and as they are needed.

This work has been designed with flexibility in mind and readers should feel free to utilise it in whatever manner they see fit. For those with existing skills or a clear vision which they wish to realise, it will provide a flexible reference guide. But for those readers who lack any pre-existing familiarity with the documentary-making process, we invite them to begin at Chapter One and follow the course of the book as it is written. Engage with the video lessons and the assigned practical exercises. By the end of that process, we believe you will have the necessary skills to realise your filmic projects on your own terms.[1]

---

1 William DeJong, Eric Knudsen, and Jerry Rothwell, *Creative Documentary Theory and Practice* (London and New York: Routledge, 2021).

1. The Humanist Auteur

Fig. 1　An open access, ten-part video series is included as a part of this text. To watch the first video lesson, readers of the online edition of this text should click on the link reported below. Readers of the print book can access the video by scanning the above QR code. Users can do this by opening the camera application on their phone and taking a photograph of the QR code. http://hdl.handle.net/20.500.12434/0322725a

Humanities scholars are frequently wary of documentaries — often with good reason. Countless documentaries produced by a range of corporate and public bodies have prioritised entertainment over factual accuracy, shock value over critical thinking, and newsworthy soundbites over a sound interpretative foundation. Over-simplification is a common problem. Academic inquiry is frequently manipulated to provide a sense of undeserved credibility. Unqualified presenters leaf through old documents and ruminate on their brilliance, claiming credit for 'new' discoveries.

© 2021 Darren R. Reid and Brett Sanders, CC BY-NC 4.0　　https://doi.org/10.11647/OBP.0255.01

Too many documentaries prioritise the desire to entertain over the need to enlighten. Their research might well be out-of-date and the conclusions they draw (often depicted as shocking or paradigm-shifting) tend to be nothing of the sort. Acts of blatant plagiarism are reframed as brilliant innovations. Dashing presenters speak with such authority that their audience can hardly begin to doubt them. Old rooms are opened for the 'first' time. Discoveries are made. Television journalists ask 'hard-hitting' questions of the qualified and unqualified alike. Fantasy is presented as reality. The humanist scholar is undermined.

These issues reflect the dangers associated with producing poor-quality or intellectually limited films — but they are not problems inherent to the medium.[1] Indeed, the democratisation of the filmmaking process, brought about by rapid and substantial changes in affordable technologies combined with the ability to achieve near instantaneous access to a global audience, presents humanist scholars with an array of new opportunities.[2] Unlike in decades past, when documentary filmmaking was, effectively, a walled garden, scholars are now in a position to take control of the medium — should they choose to do so.

If documentaries have previously served as a medium in which non-experts have held disproportionate sway, the coming of the digital documentary has the potential to reshape that paradigm.[3] For such a disruptive wave to be realised, however, humanist scholars must first proactively work towards taking control of the medium. The emphasis

---

1 Rolf Schuursma 'The Historian as Filmmaker I' and John Greenville 'The Historian and Filmmaker II' in Paul Smith (ed.), *The Historian and Film* (London and New York: Cambridge University Press, 1976), pp. 121–31 and 132–41.

2 Mike Figgis, *Digital Filmmaking. Revised Edition* (London: Faber & Faber, 2014).

3 There are many examples of documentaries that empower non-experts over experts. In the UK, one of the most prominent beneficiaries of these is Dan Snow, a broadcaster whose work as a presenter of history documentaries has allowed him — and others who follow his example — to brand themselves as historians, gaining significant sway in the public sphere, talking about a broad range of topics, regardless of their specific qualifications. For an example see Faisal J. Abbas, '"A History of Syria," Distorted by the BBC!', *Huffington Post UK*, 19 March 2013, https://www.huffingtonpost.co.uk/faisal-abbas/a-history-of-syria-distor_b_2900053.html, and 'BBC Documentary, "A History of Syria with Dan Snow", was "Biased and Inaccurate" Say Critics', *Huffington Post UK*, 17 March 2013, https://www.huffingtonpost.co.uk/2013/03/17/bbc-documentary-history-snow_n_2896575.html. For an example of Snow's broader public profile, see Adam Sherwin, 'Dan Snow: The Historian Who's Not Attached to the Past', *The Independent*, 23 October 2011, https://www.independent.co.uk/news/people/profiles/dan-snow-historian-who-s-not-attached-past-2277687.html

has now shifted — the academy is longer victim of a filmmaking process over which it has little control. With the production of digital documentaries, the onus is now on the scholar to help reshape the media landscape to better suit their goals and ideals. Passivity will accomplish nothing.[4]

The Digital Wave (and the Power It Gives Us)

Several years ago, we were lucky enough to take part in a debate on the subject of 'public history'. The resulting discussion was telling. David Starkey, a discredited British broadcaster and onetime academic historian, was mentioned several times, and, in particular the apparent sway his problematic interpretations of the past appeared to have over the general public. In the eyes of some participants, the medium as a whole seemed to be tarnished by its association with such broadcasters.[5] Others spoke of the vast power imbalances faced by scholars who agreed to participate in professional productions. The demands of a preconceived script or belligerent producers, more interested in creating entertainment than in educating their audience, were common themes. Specialised knowledge is vital, but it is not always respected or used appropriately. Scholars could hope to exert a limited degree of positive influence, but their efforts, it appeared, were

---

4   Whether or not academics use mediums such as film to shape the discourse on the past, others are willing to do so. For a sample of the rich literature dealing with the relationship between the film industry, cinema, and the past, see Pierre Sorlin, *The Film in History: Restaging the Past* (Oxford: Blackwell, 1980), Robert A. Rosenstone, *Visions of the Past: The Challenge of Film to our Idea of History* (Cambridge: Harvard University Press, 1998), and Robert A. Rosenstone, *History on Film, Film on History* (London and New York: Routledge, 2012).

5   A large part of the discourse surrounding Starkey was concerned with his recent complaint about the 'feminised' nature of history. In particular he was critical of the way in which Henry VIII 'has been absorbed by his wives', something which he linked to 'the fact that so many of the writers who write about this are women and so much of their audience is a female audience. Unhappy marriages are big box office'. Whilst Starkey possesses academic credentials, his prominent role as a television presenter provided him with high visibility to the general public. See June Purvis, 'David Starkey's History Boys', *The Guardian*, 2 April 2009, https://www.theguardian.com/commentisfree/2009/apr/02/david-starkey-henry-viii, and Stephen Adams, 'History has been "Feminised" Says David Starkey as he Launches Henry VIII Series', *The Telegraph*, 30 March 2009, https://www.telegraph.co.uk/culture/tvandradio/5077505/History-has-been-feminised-says-David-Starkey-as-he-launches-Henry-VIII-series.html

frequently in vain. The documentary medium was utterly beyond their ability to control.[6]

That is no longer the case. Film, in its varied and evolving guises, has proven itself to be a remarkably effective way of communicating complex ideas to a broad range of audiences.[7] The technology required to produce cheap and effective documentaries is now nearly ubiquitous. All that remains is to close the skill gap and to widen discussions about the ways in which visual grammars can specifically benefit humanist discourse.[8]

Scholars are not necessarily filmmakers — and vice versa. Indeed, the two skillsets, each of which requires substantial investments of time and passion, are often startlingly different. A badly written monograph can be forgiven, but a poorly researched one, which lacks the depth of inquiry demanded by the academy, no matter how well written, cannot.[9]

---

6   Despite the seemingly alien nature of this discussion, there is actually a long tradition of academic exploration of the relationship between historians and film. The introduction to the pioneering work *The Historian and Film* by Paul Smith is the logical place to begin any such investigation. See Paul Smith (ed.), *The Historian and Film* (London and New York: Cambridge University Press, 1976), pp. 1–14.

7   There is a vast literature dealing with the intellectual complexities and potential of film. As a starting point, see Robert Arnheim, *Film as Art* (Berkley and London: University of California Press, 1957), pp. 8–34. Looking beyond this, the following represent a short sample of works to be considered: Eric Rhode, *A History of Cinema from Its Origins to 1970* (London: Penguin, 1972), Mark Cousins, *The Story of Film* (London: Pavilion, 2011), Adrian Martin, *Mise En Scene and Film Style* (New York: Palgrave Macmillan, 2014), and V.F. Perkins, *Film as Film: Understanding and Judging Movies* (London: Viking, 1972).

8   Whilst it is not the purpose of this volume to be prescriptive by suggesting which subjects or themes are or are not best suited to a visual exploration, by way of an example, studies of cinema and performing art may well be an obvious beneficiary of exploration using a medium that does not require their translation into another form — writing — into which they can be made to fit imperfectly. As an example, an article by Reid about Marceline Orbes, an important comedic performer on the stage from the early twentieth century, who influenced the likes of Charlie Chaplin and Buster Keaton, had to deal with such an issue of translation: describing movement and the body without a precise visual representation to which readers could be directed. Whilst the overall discussion in the paper achieved its ultimate goal, writing was not necessarily the most elegant fit for an analysis of the power of performing arts, even if it was an adequate medium for discussion its historical (rather than its artistic) merits. See Darren R. Reid, 'Silent Film Killed the Clown: Recovering the Lost Life and Silent Film of Marceline Orbes, the Suicidal Clown of the New York Hippodrome', *The Appendix* 2:4 (2014), http://theappendix.net/issues/2014/10/silent-film-killed-the-clown

9   For an example, see Francis Paul Prucha's review of *Bury My Heart at Wounded Knee: An Indian History of the American West* by Dee Brown, in *The American Historical*

Conversely, a documentary that entertains, but which is marred by problematic intellectual elements, can nonetheless achieve widespread acclaim. Countless popular productions attest to the importance of entertainment, even as they underline much of the mainstream industry's casual disregard for accuracy or reason.[10]

This reality helps to explain the tension between humanist scholars and the film industry. One pursues a reasonable exploration of the truth based upon an in-depth and transparent engagement with the evidence. The other pursues narrative and visual beauty, or, more likely, profit or large audience numbers; the metrics of success between the academy and the film industry are vastly different. That is, of course, an over simplification but, for the purposes of this brief discussion, it at least highlights the paradigm that new technologies (and online spaces) have made obsolete. Prior to the advent of very high-quality consumer cameras, there was no realistic way for a scholar to easily produce a documentary film without making a significant financial investment in equipment, skills, crew, and supplies. Distribution was perhaps an even greater challenge — significant investment would not guarantee that one's work would, or could, be consumed by the desired audience.[11]

The digital wave has broken down those barriers. Cameras are now comparatively affordable and highly capable, whilst the maturation of the internet has opened up an array of new ways to distribute and disseminate one's work.[12] To put it bluntly, the scholar no longer has to interact with the traditional gatekeepers of the film or television industry

---

Review 77:2 (1972), 589–90. For an example of a non-academic writer retorting to such an academic critique, see Hampton Sides' Foreword to *Bury My Heart at Wounded Knee: An Indian History of the American West* (1972) by Dee Brown (New York: Henry Holt and Company, 2007), pp. xv–xx.

10 A case in point is D.W. Griffith's much discussed *The Birth of a Nation* (1915) — a huge technical and artistic achievement, 'The Birth of a Nation' was a startling racist interpretation of life in the southern United States during the post-Civil-War era of Reconstruction. Despite its deeply problematic racial themes, the film is a triumph of sentimental nostalgia, an expert demonstration of cinema's persuasive potential. As critic Roger Ebert once put it, '"The Birth of a Nation" is not a bad film because it argues for evil. Like [Leni] Riefenstahl's "The Triumph of the Will," it is a great film that argues for evil.' See Roger Ebert, 'The Birth of a Nation Movie Review (1915)' *RogerEbert.com*, 30 March 2013, http://www.rogerebert.com/reviews/great-movie-the-birth-of-a-nation-1915

11 Genevieve Jolliffe and Andrew Zinnes, *The Documentary Filmmakers Handbook* (New York: Continuum, 2006), pp. 344–82.

12 Figgis, *Digital Filmmaking*.

should they wish to create a documentary film. Profit and audience size (i.e., broad and inclusive appeal) need not play a role in the production of scholarly films — nor should technical hurdles. The technological shift away from celluloid and the rapid spread of extremely high-fidelity digital cameras has reshaped the relationship (or at least, the potential relationship) between the scholar and the documentary film.

When we gathered in 2009 to discuss a Master's degree in public history (and to debate the merits and weaknesses of our taking part in documentaries) that technological shift was not yet evident, even though there were early signs pointing to the disruptive potential of the coming digital wave. George Lucas's *Star Wars: Episode II — Attack of the Clones* heralded the industrial transition from celluloid to digital as early as 2002.[13] In 2008 the Canon5D Mark II hit the market, a DSLR (digital single lens reflex — cameras with interchangeable lenses) whose video recording quality was so high that it was used to film some episodes of the wildly popular American sitcom, *House* (2004–2012).[14] The 5D Mark II brought professional quality video recording to the market for less than $3,000. Its successor, the 5D Mark III, released in 2012, continued this trend, allowing for incredibly detailed and cinematic footage to be captured by professionals and non-professionals alike. The 5D series (one of several product ranges to bring cinematic quality to consumers) exemplified the filmic empowerment of the masses. Aside from being widely lauded and utilised by independent filmmakers, Canon 5Ds have been employed in numerous top-tier productions, including Marvel/Disney's multi-billion-dollar *Avengers* franchise.[15] For consumers, this was a stunning development. Whatever the implications for the future of camera technology in Hollywood, it was a very clear indication that

---

13   Cousins, *The Story of Film*, p. 457.
14   Vlad Savov, 'Canon 5D Mark II Used to Shoot Entire House Season Finale, Director Says "It's the Future"', *Engadget*, 13 April 2010, https://www.engadget.com/2010/04/13/canon-5d-mark-ii-used-to-shoot-entire-house-season-finale-direc
15   The Canon 5D Mark II has been used to shoot sequences, not only in independent film but in large-scale Hollywood blockbusters and big-budget serialised television. In 2010, for example, the entire finale of the Hugh Laurie series *House* was shot using the camera. In 2011, Canon announced that the 5D Mark II was used to capture footage in Marvel's *The Avengers*. 'Canon Press Release: *House*', April 2010, http://cpn.canon-europe.com/content/news/EOS_5D_mark_II_shoots_house.do and 'Canon Press Release: The Avengers', 9 May 2012, https://www.usa.canon.com/internet/portal/us/home/about/newsroom/press-releases/press-release-details/2012/20120509_avengers_pressrelease

cinematic image quality would no longer be the domain of well-funded, professional organisations alone.

For those working with even smaller budgets, non-specialised equipment has reached a quality that can, with care, allow professional-style productions to be shot by practically anybody. Virtually everyone carries a device in their pocket capable of capturing footage in at least 1080p or 4K resolution.[16] Moreover, that very same device connects its owner to the greatest global distribution model in human history.[17] Scholars are thus facing a world in which they are empowered to make films and to disseminate them to a trans-national audience, with equipment most of them already own. From a technological standpoint, at least, there is nothing to stop a determined scholar from using the equipment that is probably within six feet of them right now, in order to challenge traditional academic outputs. Whilst traditional modes of academic writing have proven themselves versatile and adept, documentaries provide new scholarly opportunities. Technology is now a facilitator, rather than a barrier.

Film as Scholarly Tool

Film is not directly comparable to academic articles or monographs. The two mediums can be used to produce work of equal weight — but they are not analogous.[18] Rather, film provides scholars with a visual language and grammar, distinct and functionally different from the written techniques and forms in which most humanist scholars are trained. It is this distinction that allows film to offer a genuine alternative to traditional academic writing. When the written word provides the most appropriate medium through which an intellectual process can be explored, it should be utilised. Equally, when a filmic visual

---

16 Tony Myers, 'Lights, Camera…iPhone? Film-Makers Turn to Smartphones', *The Guardian*, 9 February 2012, https://www.theguardian.com/technology/blog/2012/feb/09/filmmakers-turn-to-smartphones

17 For a discussion on this see director/producer Don Boyd's commentary from 2011 in which he recognised the fundamental shift that occurred around the turn of the twenty-first century's second decade (at least as far as mass participation in digital filmmaking was concerned). Don Boyd, 'We are all Filmmakers Now — and the Smith Review Must Recognise That', *The Guardian*, 25 September 2011, https://www.theguardian.com/commentisfree/2011/sep/25/all-film-makers-smith-review

18 Rosenstone, *History on Film, Film on History*, pp. 125–50.

language offers clear advantages to scholars, they should be prepared to engage with that medium. Failure to do so would necessarily reduce the effectiveness of the resultant work as it attempts — but ultimately fails — to surpass the limitations of the written form.

Roland Barthes framed the mechanisms of this opportunity in 1980. According to Barthes, a photographed image is composed of two distinct elements, the studium and the punctum. The former represents the way in which the subject of a photograph can be interpreted in a cultural or political framework — through what we might consider a scholastic lens, in other words.[19] The latter, however, is the part of the image that touches the viewer on a personal level — the subjective discourse generated by the interaction between photographer (or the filmmaker, in the context of this discussion) and their audience.[20] Understanding these two components of the photographed image allows the photographer — or critic — to understand its successes and failures, to explore the depths of the discourse, both academic and emotional, generated by the image. Something similar is true of scholars who use film. They must understand the medium's emotional, as well as its scholarly, potential.

As a medium that juxtaposes complicated visual and audio elements, often in a very controlled and time-specific manner, film offers new opportunities for scholars to explore the relationship between their work and their audience; to invite (or disinvite) emotional resonance which complements or problematises the intellectual basis of their study. A historian exploring the emotional or subjective realities of a post-war society, for instance, might well find that documentary, with its potential to simultaneously contrast different elements (and thus ideas), provides

---

19 In all likelihood, Barthes did not identify the studium as a scholarly filter. Rather, he saw the studium as the way in which a photographic image was understood by the collective — the imposed framework of the collective understanding as opposed to the more subjective understanding (punctum) each individual creates in a relation to the image. Barthes's idea, however, is adaptable and, as Michael Fried has shown, it is in need of careful deconstruction. In the context of scholarly filmmaking, the collective understanding can reasonably be re-orientated to account for a specific collective — the academy — whilst the contrasting principle of the punctum serves to account for the relationship of the work to the individual outside of a strictly academic context. See Michael Fried, 'Barthes' Punctum', *Critical Inquiry* 31 (2005), 539–74.

20 Roland Barthes, *Camera Lucida* (New York: Hill and Wang, 1981).

a distinctly satisfying method of exploring their topic. Scholars of film, music, and other performing arts might, in perhaps more obvious ways, benefit from the use of film, as it provides them with a medium that allows for the seamless integration (and reproduction) of their sources. In contrast, written works based upon the performing arts require the scholar to translate the performance into a distinctly non-native form; melody and motion can be described, but never accurately captured in this manner.[21] Film offers new opportunities for scholars to simultaneously present — and contrast — ideas, performance, and abstract interpretation.

David Mamet, the Pulitzer-prize-winning playwright and director of film, argues that the power of movies is to be found in their ability to juxtaposition one image, or set of images, against another. According to Mamet, whose ideas are rooted in those of Soviet cinematic master Sergei Eisenstein, the power of a film is not to be found in any individual image; rather it is to be found in the contrast created when one shot is placed next to another.[22] The difference, contrast, shock, or comfort of different shots, he argues, provides the emotional — even intellectual — resonance of the moment.[23] For the filmmaker-scholar, emotional or intellectual substance may be attained through the contrast between voice-over (deadpan and emotionless) versus the actual text being read (a personal self-reflection); or between the imagery on screen and the intellectual conclusion being drawn by the narrator; or, in a more directly Eisensteinian fashion, the contrast between different shots — filmic elements not running in parallel but sequentially.

Alternatively, the humanist scholar may well reject the emphasis placed by Mamet upon the juxtaposition. Instead, they might find, particularly as they gain experience with the camera, that an individual shot, not cut or otherwise substantially edited, can contain all of the necessary and desired intellectual and emotional resonance. Indeed, there is much to be said for the unflinching eye that the camera can provide. In the opening of his 2009 film, *Capitalism: A Love Story*, Michael Moore demonstrates this by showing his audience a home movie,

---

21 This was something I experienced first-hand in a analysing performing arts (see note 8).
22 Anne Nesbet, *Savage Juncture: Sergei Eisenstein and the Shape of Thinking* (London and New York: I.B. Taurus, 2003), pp. 1–20.
23 David Mamet, *On Directing* (New York: Penguin, 1992), pp. 1–7, 26–47.

filmed by a family as they are evicted after failing to keep up with their mortgage payments.[24] When taken as a whole, *Capitalism: A Love Story* is practically defined by contrast and juxtaposition. Its opening sequence, however, stands apart from the larger production, a short film within a film. Moore's commentary, which arrives after several pained minutes, does little to meaningfully deepen the power of the sequence; emotional resonance was already thoroughly accomplished with only minimal external interference. Indeed, it was the consistency of the moment, the steady perspective (if not emotional state) enabled by the footage, which mires the viewer in the family's plight. Juxtaposition would likely have served only to distract from the emotional resonance present in the original footage.

By rejecting or embracing Eisenstein and Mamet (by experimenting with and critically reading the conventions of documentary and narrative films), the humanist scholar may well find a specific filmic grammar which will allow them to explore their intellectual ideas in new ways. Such an approach does not necessitate the abandonment of traditional academic publications. Instead, it is an opportunity to broaden the tools at the scholar's disposal, to approach their subject with a new set of visual conventions (filmic grammar) that will allow them to complement a more traditional body of written work. The digital shift in the industry has now opened up the medium of film and documentary to humanist scholars — the grammar of film is now fully within their grasp.[25]

The Filmmaker-Scholar

As with any means of presenting research, using film requires the author to develop and hone a wide array of skills. This, more than anything else in the age of digital film production, is the primary barrier that separates

---

24 *Capitalism: A Love Story*. Directed by Michael Moore. Los Angeles: The Weinstein Company, 2009.

25 For discussions on the potential, and early limitations, of this technological shift see Ana Vicente, 'Documentary Viewing Platforms'; Danny Birchall, 'Online Documentary'; Patricia R. Zimmermann, 'Public Domains: Engaging Iraq through Experimental Documentary Digitalities'; and Alexandra Juhasz, 'Documentary on YouTube: The Failure of the Direct Cinema of the Slogan', in Thomas Austin and Wilma de Jong (eds), *Rethinking Documentary: New Perspectives, New Practices* (Maidenhead: Open University Press, 2008), pp. 271–77; 278–84; 285–311.

the scholar from the filmmaking process. As filmmaker Michael Rabiger once put it, 'the insights and skills required to be a minimally competent director are staggering.'[26] To produce an intellectually successful documentary is no simple task. Capturing footage is comparatively easy, but capturing *effective* footage poses significant challenges, and, once captured, assembling it into a coherent, larger piece poses yet another set of hurdles to overcome.

Acquiring the necessary documentary-making skills is a challenge, but the potential benefits are significant. In undertaking this task, the humanist scholar will gain a new vocabulary and grammar through which they can explore their ideas and research.[27] Just as learning to write in an academically rigorous and effective manner encourages thinking in a highly ordered, logical, and clear manner, the process of becoming a filmmaker provides the scholar with new ways to think through their problems.

For instance: the process of editing is, in practical terms, the art of juxtaposition — the placement of different images in adjacent chronological spaces whose contrast, established as much by the timing of the cut as the content of the individual shots, helps to shape the viewer's impression of the issue being explored. For Eisenstein and Mamet this process created the intellectual heart of their works. Their precise control over the viewed experience allows the filmmaker to carefully shape their audience's perception of an issue, not in a way that is superior to the written word but in a way that is functionally distinct.[28] In film, the scholar can precisely time images and cuts, showing a specific visual montage rather than having to make an appeal to the imagination, as writers must do of their readers. Writing invites imaginative spaces to be constructed, whereas filmmaking furnishes such spaces with pre-made images and juxtapositions. As a result, new theses, previously difficult to express in a non-visual form, might well become more achievable and more desirable.[29]

---

26 Michael Rabiger, *Directing: Film Techniques and Aesthetics. Third Edition* (London and New York: Focal Press, 2003), p. 6.
27 Christopher J. Bowen and Roy Thompson, *The Grammar of the Shot* (London and New York: Focal Press, 2013).
28 Mamet *On Directing Film*, pp. 3–7; 31–33.
29 For an introduction to how film creates these imagined spaces and, specifically, how the filmmaker-scholar can achieve their desired effect, see Greg Keast, *Shot*

In order to realise this potential, it is necessary to commit to a new learning process. Camera operation, shot framing, the psychology of cinematic photography, the theory of editing — all are necessary, but all offer new opportunities to reflect upon the nature of one's research, methodology, and intellectual dissemination.[30] As a result, the process of learning these skills enhances the scholar by bringing them into direct contact with artistic creation, bridging a gap between the arts and humanities not typically straddled in modern academia.

At a fundamental level, the arts and humanities are the same thing. Both explore the nature of human experience and our relationship to the broader cosmos; each field endeavours to encourage thought and critical discourse, to use their respective mediums to problematise and explore accepted notions; to provoke responses which, in turn, will require further discussion and analysis. Their modes of expression and their chosen mediums are vastly different but, at the most foundational level, common DNA links Da Vinci's *Mona Lisa* to Machiavelli's *The Prince*. Both are meditations on the nature of the self, albeit in very different ways, and of the relationship between the individual being and the wider world they inhabit.[31]

Documentary films produced by humanist scholars embrace, even if only unconsciously so, the link between the humanities and the arts. In that sense, the production of such films is a logical, evolutionary step

---

Psychology: The Filmmaker's Guide for Enhancing Emotion and Meaning (Honolulu: Kahala Press, 2014); Sheila Curran Bernard, *Documentary Storytelling: Creative Nonfiction on Screen* (New York and London: Focal Press, 2014); and James Quinn (ed.), *Adventures in the Lives of Others: Ethical Dilemmas in Factual Filmmaking* (New York: I.B. Taurus, 2015).

30   Michael Rabiger, *Directing the Documentary* (Abingdon: Focal Press, 1987).

31   See Joanna Woods-Marsden, 'Portrait of the Lady, 1430–1520', in David Brown Alan (ed.), *Virtue and Beauty: Leonardo's Ginevra de' Benci and Renaissance Portraits of Women* (London: Princeton University Press, 2001), pp. 64–87; Gustav Kobbé, 'The Smile of the "Mona Lisa"', *The Lotus Magazine* 8 (1916), 67–74; Kenneth Gouwens, 'Perceiving the Past: Renaissance Humanism after the "Cognitive Turn"', *The American Historical Review* 103 (1998), 55–82; Felix Gilbert, 'The Humanist Concept of the Prince and the Prince of Machiavelli', *The Journal of Modern History* 11 (1939), 449–83; Charles D. Tarlton, 'The Symbolism of Redemption and the Exorcism of Fortune in Machiavelli's *The Prince*', *The Review of Politics* 30 (1968), 323–48; Joseph D. Falvo, 'Nature and Art in Machiavelli's The Prince', *Italica* 66 (1989), 323–32; Victoria Kahn, '*Virtù* and the Example of Agathocles in Machiavelli's *Prince*' *Representations*' 13 (1986), 63–83.

in an increasingly digital, creatively egalitarian world.³² Indeed, the scholarly production of documentaries is a post-digital process in the sense that it marries the digital (new technologies) to the analogue (real world interactions). The relationship between the self and society — and the relationship of both to the wider cosmos — remains the main focus of the humanities, but documentary-making provides an opportunity to explore those issues in a way that transcends disciplines. The humanists' new tool is artistic expression.³³

In that sense, the scholar is enhanced when they embrace new technologies that allow them to step outside the traditional parameters of their subject area. The construction of a film requires not only the fostering of new skills, but a reflection upon the ways in which the discussions typically explored by scholars using written language can be transferred to a medium that is primarily visual in nature. Documentary films are often wildly different from one another, providing scholars

---

32 Jeremy Harris Lipschultz, *Social Media Communication: Concepts, Practices, Data, Law, and Ethics* (New York: Routledge, 2015).

33 Rosi Braidotti, the post-humanist thinker, has argued that the future of the humanities lies in the crossing of disciplinary lines and the exploration of subject areas not traditionally linked to the humanities. According to Braidotti, the changing nature of the human experience will necessitate changes in the humanities which will, according to her, require further trans-disciplinary interaction. This prediction is bold — there is logic to it, but that logic leaves significant room for debate; not the least of which concerns the shape of future trans-disciplinary approaches to studying the human being. Far from radical, the use of new digital technologies to facilitate the creation and dissemination of non-traditional research outputs is, in the context of Braidotti and other post-humanist thinkers, a rather modest innovation. The point being made here is not that historians and humanist scholars should try something that is (in the purest sense of the word) new. Rather, they should instead try something that has its ideological and intellectual precedent in the trans-disciplinary world of the Renaissance. The production of digital documentaries is, in that sense, simultaneously new and old. New for most humanist scholars but, at a base intellectual level, perfectly consistent with the trans-disciplinary spirit of our humanist and Renaissance-era antecedents. The process of scholarly documentary-making, then, is one that is utterly facilitated by the emergence of new digital tech — but is linked to centuries-old ideas in which disciplinary boundaries are seen as malleable. Taken to its natural conclusion, disciplinary boundaries must melt away in the face of scholarly investigations into the nature of the human being and the dissemination of that knowledge. Specialisation in this model is less about specialisation within a traditional field than it is with specialisation in a concern for the broader human experience, and the need to utilise whatever fields or approaches allow for the study (and dissemination) of complex and enlightening potential truths. For a further discussion on these ideas, see Rosi Braidotti, *The Post-Human* (Cambridge: Polity Press, 2013), pp. 143–85.

with a significant degree of freedom to experiment.[34] There is no standard template for a scholarly documentary beyond that which their authors are able to define.

The transition from written pieces to cinematic ones can create practical problems, to be sure. References, for instance, are not easily integrated into the documentary medium. There are, however, a number of potential solutions that can be employed to overcome some of the hurdles presented by a new scholarly medium. A written appendix containing references or methodological discussions would be a clumsy, though effective, solution to the referencing dilemma. A more innovative approach might be the addition of interactive elements to the film, such as a small icon that appears whenever a reference or footnote is required, which provides the viewer with the option of bringing up the relevant information.[35]

More problematic for the filmmaker-scholar may be their belief (likely fuelled by preconceived ideas) that they should strive to create films that entertain as much as they enlighten — but this is only a consideration if the plaudits of traditional film critics and audiences are desired. There is no reason for a scholar to suspect that the production of a documentary film will lead to a vulgar expression of their ideas; it is their medium to (re)define as they see fit. Indeed, scholars should be willing and eager to challenge convention. After more than a century of intensive development and refinement, the mainstream film industry has honed a number of well-realised formulas — a schema that is instantly recognisable as a satisfying or entertaining experience.[36]

---

34 Consider, for instance, Robert J. Flaherty's 1922 film *Nanook of the North*, which fictionalised and staged much of its content, but which nonetheless succeeds in creating a narrative that brought Alaskan aboriginal peoples, even if a fictive version of them, into the mainstream culture. Then consider Neil Diamond's 2009 film *Reel Injun* which explores the long-term damage of the so-called 'mainstreamification' of aboriginal cultures. Both are so vividly different as hardly to merit comparisons — and yet they are also similar in both form and content; so much so that, when taken together, a new narrative of aboriginal empowerment in the Americas begins to emerge. See *Nanook of the North*. Directed by Robert J. Flaherty. New York: The Criterion Collection, 1999 and *Reel Injun*. Directed by Neil Diamond. Montreal: National Film Board of Canada, 2009.

35 Dayna Galloway, Kenneth B. McAlpine, and Paul Harris, 'From Michael Moore to JFK Reloaded: Towards a Working Model of Interactive Documentary', *Journal of Media Practice* 8 (2007), 325–39.

36 According to Bill Nichols, documentary can exist in one of six forms — the poetic, expository, participatory, observational, reflexive, or performance. For a discussion on the forms of documentary films, see Bill Nichols' discussion on his construction

There is, however, nothing to stop humanist scholars from challenging audience expectations by subverting or reimagining this model.

Embracing documentary film as a means of disseminating research does not necessarily require scholars to embrace the mainstream, or even to seek a broad audience. The scholar remains free to challenge existing conceptions and constructs.

The Filmmaker-Scholar as Auteur

If mainstream documentaries fail to offer the type of insights, deep analysis, and discussions that academic scholars find valuable, reliable, or even ethically tolerable, it is the lack of scholarly oversight and control that is to blame. In mainstream documentaries, the scholar is all too often an advisor or spectator. As a result, documentaries are developed to suit the agenda of filmmakers (and their financiers) rather than the academy. Largely absent is the *scholar-auteur* — the filmmaker-scholar with complete creative control over a film, whose influence is felt in every aspect of the production. The coming of the digital wave and its resultant democratisation of the filmmaking and distribution processes offers the opportunity for scholars to empower themselves. Whilst the traditional mainstream documentary, and its associated and problematic relationship with the academy, is unlikely to disappear in the near future, scholars are no longer powerless. They can challenge the mainstream. Indeed, considering the exploitative nature of some documentaries (see The History Channel's *Ancient Aliens* (2010-present)) they may even have a moral obligation to do so.[37]

At its most fundamental level, auteur theory argues that a film is, effectively, the creative vision of one person (or small group) whose ideas define the finished piece. One vision, one author, in other words.

---

of the Documentary Mode in Bill Nichols, *Introduction to Documentary Film* (Bloomington and Indianapolis: Indian University Press, 2001), pp. 99–137.

37 The *Ancient Aliens* example is not a flippant aside. Many problematic productions have been created by and for companies such as the History Channel — they are certainly not unique in that regard. And though the reader of this volume might safely be assumed to pay series such as *Ancient Aliens* little heed, there is an audience who trusts programs such as this and, partly thanks to the professionalism of those productions, consider their arguments and evidence to be a valid candidate for the truth. Such audiences should not be looked down upon by the academy — nor should they be ignored or abandoned.

According to this theory, which de-emphasises the implied collaboration between every member of a production, through active agency or passive endorsement, films must necessarily represent the specific and focused desires of their chief creator, the auteur. Authorship of films is precise and attributable; the creative zeitgeist is thus linked inextricably to a core creative talent.[38]

Setting aside debates about the universal veracity of the idea, auteur theory provides an excellent framework with which humanist scholars can begin to conceptualise their role in the emerging media landscape of the digital era. As invited participants and advisors, the humanist scholar's influence over documentary production tends to be limited. Well-honed arguments and careful research no doubt impact many productions but, fundamentally, a lack of direct creative control can only serve to disempower the humanist scholar. In the face of a strong-willed producer or director, no matter how ill-informed they may be, the humanist scholar has little power of enforcement and, though it may be loathsome to admit it, a compelling argument does not necessarily win the day. The scholar can, of course, attempt to exert positive change over the productions in which they are involved — but they cannot enforce their beliefs. More problematic still is the far larger body of scholars who are not invited to participate in such productions at all, whose research and perspectives are therefore completely excluded from the conversation. Far from serving as auteurs, scholars tend to be marginalised — used when they are perceived to be of value, but just as likely to be ignored.

The scholar-auteur, then, tends to be conspicuous through their absence. This is the paradigm that the digitisation of the filmmaking process, and the democratisation of distribution channels, allows the academy to challenge. Properly motivated, and willing to develop the necessary skills, there are few reasons why humanist scholars cannot take the place of the director or producer, to develop a creative — or rather, intellectual — vision which is reflected in every part of a finished production. Research, argument, deconstruction, logic, and visual grammar can all be controlled directly by the filmmaker-scholar. In so doing, they will take control of a mode of academic expression that is often

---

38 Andrew Sarris, *You Ain't Heard Nothing Yet: The American Talking Film, History and Memory, 1927–49* (Oxford: Oxford University Press, 1998).

controlled by those outside of the academy; through experimentation and imagination, they will be able to realise a visualisation of their intellectual vision rather than aiding the outsider in realising theirs. The filmmaker-scholar will become the scholar-auteur.

A willingness to engage with the medium and to experiment will allow scholars to challenge and exploit it; to create opportunities to present primary evidence in new ways; to juxtapose and explore ideas visually; to reach specific audiences, broad and niche; to generate an audience-based feedback loop through the interactive nature of modern distribution channels, which solicit comment and generate online discussion; to engage in multi-perspective subjective explorations of thesis and concept. A self-conscious decision will need to be made to facilitate this — not a willingness to participate in mainstream documentaries when invited, but a desire to proactively take control of the medium by mastering every aspect of the production process (or forming a team with the required range of skills). Auteur-ism should be recognised — and embraced.

With direct creative and intellectual control of a documentary project the scholar will face challenges, not the least of which will be securing the resources necessary to create a high-quality documentary output. Aside from the intellectual resources in question — the baseline skills, which can and will be learned through study and practice — more material concerns will prove to be an issue. As with the independent film movement, however, the scholar-auteur will overcome these limitations through imagination and the intelligent deployment of the resources available to them. By learning a wide array of skills, from camera operation to sound recording and editing, the need for a crew will be reduced — or even eliminated. Engagement with students and other scholars in new pedagogical and collaborative spaces is one possible avenue to overcoming this deficit if complete self-sufficiency is neither possible nor desired. The careful use and management of existing and available resources — the planning of production around what is easily available to the filmmaker-scholar — will facilitate academic engagement with the documentary medium.

The filmmaker-scholar can benefit from the immense amount of material produced by independent and mainstream filmmakers. A wide corpus on the theory and practice of film production is readily available — and independent filmmakers, through their writing and work, continually demonstrate how new technologies, techniques, and imagination provide solutions that can facilitate the work of the scholar-auteur. As a result, they demonstrate that the barriers of even the recent past have been demolished. The use of documentary film as a means of disseminating research and engaging in intellectual discourse is now within the hands of the scholar.

The filmmaker-scholar, as imagined in this book, is a scholar who sets aside any negative, preconceived ideas that they might harbour about documentary films. They do not recognise the form as being limited, a way to communicate with a mass audience via twentieth-century staples such as television, but instead celebrate the unique opportunities that a complicated layering of audio-visual elements offers them. They recognise that the documentary is a malleable form, which has been affected by disruptive changes brought about due to the emergence and proliferation of new technologies. They may well aspire to produce films that are projected on large cinema screens, or they may envision their works being consumed primarily on smartphones. Either way, they will recognise, identify, and attempt to exploit the potential of the medium to explore their intellectual ideas and research in new and intriguing ways.

The filmmaker-scholar rejects the idea that the academy cannot be in control of the documentaries that are consumed by broad and niche audiences alike. They do not wait for traditional gatekeepers of the medium to invite their participation, nor do they accept that they cannot possess complete creative control of a production. The filmmaker-scholar may well participate in the projects of others, but they create projects of their own, developing and realising their intellectual and creative vision. Their films reflect these visions, presenting candidates for the truth that are rooted in their research and intellect. The filmmaker-scholar cannot deflect the blame for an unsuccessful project — in a very real sense, they are its author.

Documentary film presents opportunities to expand discourses within and without the academy, a reality the humanist-auteur recognises and celebrates. They embrace academic forms of publication beyond the monograph-article dichotomy, which they may still employ, perhaps even as their principal avenue for publication. The humanist-auteur will be no less dedicated to academic and scholastic excellence than their peers. Whether through book, film, or journal article, the humanist-auteur's first loyalty will be to the creation of reasoned analysis disseminated through the most appropriate form (written, filmed, or otherwise) which is available to them.

Looking for Charlie

Fig. 2 Watch Looking for Charlie by clicking on the link below or scanning the QR code. *Looking for Charlie: Life and Death in the Silent Era*. Digital Stream. Directed by Darren R. Reid and Brett Sanders. Coventry: Studio Académé, 2018. http://www.darrenreidhistory.co.uk/stream-looking-for-charlie/

As an example of what an ambitious documentary *might* look like, we present to you our feature film debut — *Looking for Charlie: Life and Death in the Silent Era* (2018).[39] You can stream the film for free by pressing the play icon in the embedded video above or by scanning the QR code (if you are reading the print edition of this book).

Looking for Charlie was a very ambitious project. It took us three years to make and was shot principally in New York, London, Nuremberg, and Hong Kong. It is an in-depth examination of life in the silent era, focusing upon the hidden figures who helped to shape iconic performers like

---

39 *Looking for Charlie: Life and Death in the Silent Era*. Directed by Darren R. Reid and Brett Sanders. Coventry: Studio Académé, 2018.

Charlie Chaplin and Buster Keaton. But it is also an examination of the role played by mental health in this era; two of the hidden figures in the film took their own lives, whilst Chaplin and Keaton had mental health issues of their own. As the project progressed, we recognized that there was a lot of overlap between our own experiences with mental health and those of our subjects. We thus chose to integrate own experiences a part of the film's larger narrative. In other words, *Looking for Charlie* is a thoroughly personal, idiosyncratic project in which subjective reflections sit next to more intellectual observations and analysis. It is a project that embraced the auteur-ish possibilities of the medium.

Traditional academic writing has few spaces for such deep, subjective engagement.[40] The documentary medium, however, with its different expectations and rather undefined place within the academy, offered us an opportunity to explore our topic in an open, personal, and constructive manner. You are under no obligation to follow a schema similar to our own. *Looking for Charlie* is not presented here as a blueprint; only as an illustrative example for readers to enjoy, reject, build-upon, react against, or ignore entirely.

Academic documentaries can be an extension of existing scholarship; a conduit through which scholars can reach a broad (non-scholarly) audience; and they can become something else entirely. With *Looking for Charlie* we erred towards the latter, not because we felt that all academic documentaries should engage in personal, subjective reflection, but because such an approach ultimately satisfied the intellectual and emotional goals of this particular project.

Your goals, personality, and intellectual framework will no doubt differ from our own. This may lead you to create radically different works from our own. We embrace that diversity of perspective.

---

40 For an example of some element of the reflective-self appearing in an academic text, see Christopher Leslie Brown "Foreword" in Winthrop D. Jordan, *White Over Black: American Attitudes Towards the Negro, 1550–1812. Second Edition* (2012; Chapel Hill: University of North Carolina Press, 1969), pp. vii–xvi.

2. Learning to Love the Camera

It is worth taking a moment to reflect and take stock. Making a film, of any length or complexity, is a wonderful experience, filled with unique and thoroughly satisfying challenges. In some ways, we are all filmmakers.[1] Perhaps the most important footage shot this century was that which captured the planes flying into the World Trade Centre. To be sure it was badly framed, the resolution was low, and the camera shake is almost unbearable. But those short pieces of film are far more important than any of the $100+ million blockbusters that have followed since. Long after Michael Bay's *Transformers* movies are relegated to the memories of a few elderly millennials, scholars and the public will still look to that shocking footage, unintentional masterpieces of the moment, and gasp in horror.[2]

The relative crudity of such footage does not reduce its effectiveness. The footage of Rodney King's beating at the hands of the LAPD, captured on a consumer camcorder by an outside observer, is a dispatch from the frontline.[3] It is far more emotionally affective than most staged pieces that aim to produce a similar effect. It is the honesty of that footage that gives it power.[4] In all likelihood, there is footage on your phone or computer right now that is more honest and meaningful (at least to you) than anything you will see at the multiplex this year. There are moments of beauty, located on the very same device that you use

---

1. Don Boyd, 'We are all Filmmakers Now — and the Smith Review Must Recognise That', *The Guardian*, 25 September 2011, https://www.theguardian.com/commentisfree/2011/sep/25/all-film-makers-smith-review
2. For a discussion of the ways that events like 9/11 have shaped and challenged the dominant schema, see Jacqueline Brady, 'Cultivating Critical Eyes: Teaching 9/11 Through Video and Cinema', *Cinema Journal* 42 (2004), 96–99.
3. George Holliday, *Rodney King Tape*. Camcorder footage. Los Angeles, 1991.
4. For a discussion on the impact of the King beating, see Ronald N. Jacobs, 'Civil Society and Crisis: Culture, Discourse, and the Rodney King Beating', *American Journal of Sociology* 101 (1996), 1238–72.

to order your groceries. That your smartphone serves many purposes, many of them banal, does not diminish the truth or power of the scenes you have captured with it.

Most people probably do not consider themselves to be filmmakers, but almost all of us make films. They may be crudely shot, badly framed, isolated moments with no narrative or innate beauty evident to outside observers — but they are important. There have been many occasions in the history of cinema where filmmakers, from Andy Warhol to the Italian Neo-Realists, have deliberately fostered such crudeness.[5]

You are already a filmmaker — and yet you are nothing of the sort. You document your own life (and the lives of those around you), but you do not capture the types of films that people outside of your immediate social circle would likely appreciate. You were already a filmmaker — but now you have chosen to be a *deliberate* filmmaker. You want to consider your shots, cut different pieces of footage together, and create something that is important to people beyond your immediate acquaintances. The change that you wish to make is attitudinal. Start thinking like a filmmaker: how can the skills, motivations, and experience you already possess be used to impact a broader audience?

Please fetch your camera.

It does not matter if it is the phone in your pocket, just pick it up and hold it. Observe its lines with your eye, noting the different materials out of which it is made. Observe the size of its lens. Is it a large, belonging to a DSLR? Or is it small and compact, the lens of a smartphone? Whatever it is, observe it and appreciate it. If your camera has a lens that you can use to zoom in, play with that feature. How quickly does the lens zoom in and out? How long does it take to lock its focus?

Start by getting to know your camera and appreciating its existence. Thank it for all of the good service it has done you in the past, the innumerable moments it has preserved already, or will likely preserve in the future. When you are old and wheelchair-bound, young interlocutors will be fascinated when you show them a picture of you as you are now. Thank your camera for saving you, just as you are right now, warts and lines and wrinkles and all your beautiful imperfections. Take a picture right now to commemorate the moment.

---

[5] Greg Pierce and Gus Van Sant, *Andy Warhol's The Chelsea Girls* (New York: Distributed Art Publishers, 2018) and Vincent F. Rocchio, *Cinema of Anxiety: A Psychoanalysis of Italian Neorealism* (Austin: University of Texas Press, 1999).

We shall wait while you do.

The moment has passed. It is dead and gone and will never be again. We hope you captured it. Do not ever forget that every moment you ever experience will ultimately be lost (as Rutger Hauer so eloquently put it) 'like tears in the rain'.[6]

Your camera is a powerful device and, over the course of this book, you are going to learn how to harness that power as effectively as possible. That process begins by appreciating what you have right now. You almost certainly have a device that will allow you to capture a fidelity of footage that would have been unimaginable to most people just fifteen years ago. And you are uniquely you — the only person exactly like you, with a unique perspective, set of life experiences, and future. And even that will change. In a few years, the person you are now will be gone. A memory will remain, but the current entity bearing your name will be replaced by someone else, someone whose life and experiences have changed them, maybe for the better, maybe for the worse. Either way, they will be changed.

Consider the implications of that for a moment. You have two lenses through which you are going to see the world as you work through this book. The lens in your hand, your camera, and the rather more abstract lens through which you currently experience, perceive, and interact with the world. It is both of those lenses, working in tandem, that give a filmmaker their power — one lens helps to focus the other. The intellect identifies a subject worth shooting, the camera accomplishes that goal.

So, appreciate your camera. Take care of it. See it as an extension of yourself. Clean it. Do not let dust or other debris build up on it. If it is a smartphone, bundle all of your photography apps into one location. Experiment with its different settings and possibilities.

And use it. You do not need a reason. Pick it up. If it is cold outside, put on a coat. If it is raining, take an umbrella. Walk out of the house and take some photographs or, if you're feeling adventurous, shoot a few minutes of footage. You do not need an excuse to use it. Aim it at an interesting building, where the lines of the structure do not align quite as they should, or where the lighting hits it just so. Try changing your position. Why shoot everything from the same height? Drop onto your haunches and shoot low. Now lift the camera over your head and shoot

---

6   *Blade Runner*. Directed by Ridley Scott. Los Angeles: Warner Bros, 1982.

high. What happens when you shoot a light source? If you like cats, take a picture of the stray that walks up and down your street. It will not let you get close? Drop down low and take a picture of it at a distance, but be sure to capture the cat's surroundings, its context. Tell the cat's story in a single image.

Whatever you do, just appreciate the fun (and absurdity) of the moment. You are documenting the world around you in an instant that will never come again. Your camera is facilitating that process. So, get to know it, treat it with respect.

This may feel foolish, but if you make the decision to start treating your camera seriously, you are making the conscious decision to start thinking like a filmmaker. Treat yourself and your equipment with respect and you will have crossed the first threshold.

We are all filmmakers. The difference is that you now know it. Congratulations.

3. The Production Process

Creating a documentary, be it feature-length or short-form, can be intensely intimidating at the outset. The sheer amount of passion and dedication can leave even the most well-intentioned project unfinished or abandoned. To avoid this, you should aim to control the process as much as possible, lest it take control of you. Despite the distance between initial conception and the release of a final piece, every part of the process can be controlled and broken down into manageable segments.

Broadly speaking, the production process consists of three distinct phases: pre-production, production, and post-production. These three phases represent the planning, shooting, and assembly of your film. No one part of the process is more or less important than any of the others, because if any one stage is faulty, it can result in the failure of your project. Each stage of the process has its own inherent challenges, but by thinking about the production process in discrete stages it can be more easily managed and controlled.

The production processes behind different projects may vary considerably — every filmmaker will develop their own individual methodology. Pre-production for a highly scripted project will likely be one of the most important and intellectually rigorous stages in the whole process. For an observational documentary, however, one that follows a subject and cannot account for that subject's actions beforehand, pre-production will be more about planning logistics than fostering a very detailed vision of your final product.

Ensure that you understand what each stage of the production requires and involves. That will allow you successfully to manage the workload required to transform an idea into a finished product, ready for distribution and dissemination.[1]

---

1   For a broader overview of the production process, see Francis Glebas, *Directing the Story: Professional Storytelling and Storyboarding Techniques for Live Action and*

Pre-Production

Pre-production is the period of planning that occurs before the cameras start rolling. It is during pre-production that you, as much as possible, plan the events and processes that will need to occur in order for you to achieve your vision. If you wish to shoot in more than one location, plan out when, where, and how you will get to that location. List all of the equipment you will require. If overnight accommodation is required, investigate costs, and availability. By the end of the pre-production process, all of your logistics should be resolved. Having to book last minute accommodation during the production phase will detract from your ability to immerse yourself in the more creative parts of the process. The more you make the most of your time in pre-production, the more you will be able to achieve once production actually begins.

Pre-production involves a lot of planning, but it is also a highly creative process. It is during this stage that you conceptualise your film and plan out how you will achieve your vision. If you envision a highly scripted, pre-planned TV-style history documentary, it is during pre-production that you will write the script and plot your production schedule. If, on the other hand, you intend to create a film that is more observational or reactive in nature (perhaps involving the collection of a significant number of interviews from which a main thesis or theme will be generated), you may instead spend pre-production securing interview candidates, writing questions for them to answer, and so on. Even a film that is reactive in nature, however, should have a creative element to the pre-production phase. Imagine the types of shots you wish to achieve, how your subjects will be interviewed (sat down in stable locations or moving through spaces) and practice using your equipment with test subjects to ensure that, when the time comes, you can realise your vision. Pre-production is the time during which you prepare; prepare yourself, your script (if applicable), your crew, your camera skills, your schedule, your storyboards, etc. Plan everything that is within your power. This will ensure that when you do step out

---

Animation (New York and London: Focal Press, 2009); Michael Rabiger, *Directing the Documentary* (Abingdon: Focal Press, 1987); David K. Irving and Peter W. Rea, *Producing and Directing Short Film and Video. Fifth Edition* (Burlington: Focal Press, 2015); and Michael Rabiger, *Directing: Film Techniques and Aesthetics. Third Edition* (Burlington: Focal Press, 2003).

with your camera, you will be able to devote all of your creative and intellectual energy to the actual making of your project.

It is also a good idea to record as many of your thoughts as possible, keeping a record to draw on for inspiration at a later date. Sketching or writing out ideas will help you to visualise them. Purchase a small notebook, something dissimilar to those you normally use in your everyday life, and dedicate it to your film. Carry this 'idea-pad' with you everywhere and whenever an idea occurs, record it. If you watch a film (documentary or otherwise) and something catches your eye, take notes so you can refer back to it at another time — whether it be an interesting transition, curious use of music or sound, or even the way in which written words appear on the screen.

Fig. 3. The location titles in *Looking for Charlie* (seen here) pay homage to the caption style utilised in Marvel's *Captain America: Civil War* (2016). *Looking for Charlie* (00:25:38–00:25:46).

Our film, *Looking for Charlie: Life and Death in the Silent Era* (2018), has virtually nothing in common with Marvel's *Captain America: Civil War* (2016) — except for the large, almost full-screen text used to describe locations in both films. As *Looking for Charlie* took place around the world, much like the third *Captain America* film, we were inspired by the clarity of that film's screen-dominating captions. They were bold, novel (at the time), and communicated the changing locations of the film with exceptional clarity. So we borrowed them. We recorded sketches of how they might look in our 'idea-pad'. Always keep one eye on precedent (see chapter five) — be prepared to respond to filmic grammar, modern and historical.

Practically everything we learned as we were working on *Looking for Charlie* found its way into that 'idea-pad'. There is a page listing about a dozen possible titles for that film, still photographs, maps of New York (where we shot much of the film), questions that we might ask potential interviewees, ideas for the editing process, and (evolving) reflections on the nature of the film we were making. There are sketches for potential shots as well as discussions about the intellectual and emotional roles that certain shots might play. There are also pages and pages of notes on camera settings. Everything we needed, from practical reminders to sources for inspiration and precedent, was contained in that pad.

Pre-production is also the phase in which the realities of a shoot — including identifying key filming locations, transportation, costs, crew organisation, and so on — are organised: ideas must be turned into actionable milestones. Plan as much as you can. If you have a scripted segment, sketch out every shot and assemble a storyboard if required. If you wish to create a complicated, multi-camera sequence, plan out camera placement, calculate whether you will need assistance (a crew) to accomplish that task. Consider the time when you (and they) will be shooting. If necessary, organise transportation and meals accordingly. Build redundancies into your planning to accommodate unexpected calamity. The more in-depth the planning, the more effective your shoot is likely to be.

It is important that you regularly assess the achievability of your project (see chapter four). Documentaries are not necessarily more labour-intensive to produce than monographs or articles. If, however, you envision creating re-enactments or other complex set pieces, this may change and you will need to spend considerable time working out the nature of your collaboration with others (see chapter five) as well as the logistics which accompany such activities (food, safety, comfort, access to bathrooms, etc). Even a solo shoot, involving only the director (armed with a camera), requires such logistical consideration. You do not wish to find yourself capturing footage of an event only to discover that you do not have access to a bathroom or food. The organization of such logistics is beyond the scope of this text, but it should be something you consider as you plan your project. By doing so, you can help to ensure that the actual production runs smoothly, allowing you to focus your energies on the task at hand. The more you plan for in pre-production, the more fruitful and enjoyable any on-location work will ultimately be.

By the end of your pre-production process you should have accomplished two things. Firstly, you should have a clear idea about the type of film you want to make — your vision. Secondly, you should have a plan in place for how you intend to realise that vision, including locations you must visit, any interviews you wish to carry out, and a detailed scenario to accomplish complicated sequences or shots. Your original vision will, at least implicitly, speak to your plan. If your planning suggests an over-complicated or unachievable production process, your vision may need to be revisited. Ask yourself a simple question — does your objective justify the resources and effort required to achieve it? Revisit and revise your production plan as many times as necessary to develop a schedule of activity with which you are comfortable — and whose milestones are demonstrably achievable.

Production

Following the planning phase of your project, production proper can begin. Production is the phase wherein you set out to capture the footage, interviews, and so on, which will form the backbone of your film; the plan from your pre-production phase will thus be set into motion. For a number of reasons, most of which are no doubt obvious to you, this is the most intimidating and, often, challenging part of the entire process. The theoretical becomes real and the pressures placed upon the filmmaker can be vast. It is one thing to conceptualise a film, it is another to bring it into being.

It is crucial, then, that you have faith in yourself and your project throughout production. Understand that some things will likely go wrong. Accept this as a reality and be prepared to be flexible should a setback occur. An intended sequence may need to be abandoned; an overly ambitious plan may need to be overhauled or simplified — in such situations, stress and worry will be the result. Should this occur (and it almost certainly will) understand that it is simply part of the process. Deal with it as best you can and do not be afraid to take a step back and reassess. There is much to be said for taking a short break, sleeping on a crisis, and discovering new solutions to your production problems.

Successful production processes are about actioning your pre-production plans, and then rolling with the resultant punches. If you are a first-time filmmaker, or working with an inexperienced crew, you should certainly build redundancies into your schedule. This will give you time to finish sequences that overrun or allow you to compensate for unforeseen disasters that may affect your schedule. Interview subjects can cancel, trains can be delayed, and patience can wear thin. None of this is particularly pleasant, but neither is it easily avoidable. Build a schedule that recognises this.

However committed you are to realising your vision, never forget that the real world has as much say about the success of your production as you do. Inclement weather might disrupt your plans. A good pre-production plan will help to mitigate this, but in some situations the unforeseen will occur and leave you with few options. Rest assured, in the case of such an eventuality, you will overcome, so long as you are prepared to adapt and think on your feet. These challenges may seem daunting but remember that you are embarking on this undertaking for a reason. The intellectual and creative rewards are significant and by persevering through them and turning them to your advantage, your work will ultimately become stronger as a result. Shooting material for documentaries can be a challenge. But it also incredibly rewarding.

There is much to be said for taking time to reconsider your position: endeavour to achieve something valuable in the face of whatever challenges you encounter. Sudden changes in the weather might be frustrating, but they might also provide you with an opportunity to turn the camera on yourself and your crew. You may not have planned on a moment of introspection in your film, but the sudden change of conditions may well provide you with an unforeseen opportunity to improvise and add an extra dimension to your project. Perhaps the sudden downpour will allow you to add a moment of brief levity to your film, to break the fourth wall and to reflect upon the filmmaking process (and nature's ability to disrupt it). The unforeseen breeds creative opportunity.

If a sequence is rendered impossible by circumstance, reassess its importance. What was it meant to achieve? How might that same theme be explored in a different, more achievable way? It may be disappointing that your original vision could not be achieved, but something just as effective might be possible using the resources and conditions which are available to you. In other words, try not to get caught up in the disappointment of the moment. Accept the challenge that has been presented to you and adapt accordingly.[2]

When making *Looking for Charlie*, our original plan to shoot a moving conversation on Broadway was abandoned due to concerns that the sequence was a) too complicated and b) the desired location would be too busy. The result was a period of reassessment. Following some reflection, we agreed that some attempt at the sequence should be made but that the location should be altered to minimise pedestrian foot traffic. Whilst the original Broadway location would have provided visual beauty and symbolic significance, an alternative location (which was just as symbolic, albeit in a different way) was chosen. Though less visually beautiful, the new location allowed for multiple takes to be attempted whilst its proximity to the crew's hotel reduced many of the logistical issues.

Despite the complexity of the sequence, which featured no less than three moving cameras, two moving subjects, a roaming boom mic operator, and a support crew, all of whom needed to move in a coordinated, choreographed manner, we believed we had an achievable plan. Despite the difficulties in capturing the sequence, our crew rose admirably to the challenge. Reassessment, adjustment, and an unflinching desire to realise an achievable goal allowed us to capture a visually dynamic sequence in which we had a lot of faith.

Perhaps the sequence should never have been attempted — it was certainly ambitious. But ambition is no bad thing and, had the sequence not been successful, a simpler version could have been attempted at a later time. By thinking of camera positions and choreography in advance,

---

2   There is value in continually engaging with filmmaking literature throughout the production process. There are many works that can help to inspire you as they articulate the challenges (and solutions) that productions have had to deal with. Among some of the best examples are Mike Figgis, *Digital Filmmaking. Revised Edition* (London: Faber & Faber, 2014); Francis Ford Coppola, *Live Cinema and its Techniques* (New York and London: Liverlight, 2017); and David Mamet, *On Directing* (New York: Penguin, 1992).

Fig. 4. Walking through downtown Manhattan at night. This sequence in *Looking for Charlie* required three moving cameras to follow two moving subjects, both of which were wired for sound, whilst a boom mic operator recorded the city ambience. This was not an easy sequence to shoot, but the result was visually dynamic, taking advantage of the naturally high production values that New York offers. *Looking for Charlie* (0:30:58–0:32:37).

we were in a position to make a realistic effort to realise the sequence. The result was a kinetic, moving conversation through a bustling New York street in the dark of night. Ambition can pay off, but you will need to accept that it will not always do so. That is simply the nature of the process.³

When shooting *Aftermath: A Portrait of a Nation Divided* (2016), there was some discussion between ourselves as to whether or not it was worth shooting in the New York borough of Harlem. A prior attempt to do so had not gone according to plan due to inclement weather. Despite

---

3   There is also value to consulting works that offer cinematographic inspiration, illustrating interesting camera angles, shots, and camera movements which you might want to employ during a shoot. Some examples of such works include Gustavo Mercado, *The Filmmaker's Eye: Learning (and Breaking) the Rules of Cinematic Composition* (New York and London: Focal Press, 2010); Steve Katz, *Film Directing: Shot by Shot* (Michigan: Michael Wiese, 1991); and Jennifer Van Sijll, *Cinematic Storytelling: The 100 Most Powerful Film Conventions Every Filmmaker Must Know* (Michigan: Michael Wiese, 2005).

some reluctance to repeat the experience, we nonetheless recommitted to the locale in the hope that it would produce dynamic and arresting interview material. Harlem did not disappoint and, in our afternoon there, we were able to collect a wide array of interviews, each of which made it into the final cut of that project. It is impossible to imagine the project being a success had we not taken the opportunity to shoot there.

On the other hand, if you can capture your vision in a reasonable timeframe, using the resources you have to hand, consider taking a more ambitious path. It will likely take a significant investment of time in order to achieve a more ambitious goal, but if you have time and patience to spare, the results, though more exhausting, can add significant value to your project. Do not give up on an ambitious idea straight away, but at the same time, do not invest more resources in something unlikely to provide a significant intellectual or creative return. Do invest in those moments that you believe are achievable and that will add significant aesthetic or intellectual value to your project.

It is also worth mentioning that you should develop a rigorous end-of-day process, which will include time to care for your equipment, recharge batteries, and back up data. Every night you should ensure that all camera batteries are recharged. Memory cards should be downloaded on to at least two separate hard drives (in case one fails), and your footage should be reviewed to ensure that the material you captured meets your requirements (all your shots should be in focus, etc). This part of the daily process is non-negotiable. It is easy to lose footage and potentially very difficult, if not impossible, to capture it again. The footage you capture is the currency of your shoot and should be treated as such.

Post-Production

Considering the amount of effort expended on planning and, then, shooting your film, one might imagine that post-production would be comparatively straightforward: the assembly of your pre-made filmic pieces into a pre-determined order. In many ways, however, the commencement of post-production signals the start of a new creative phase, which is as involved as anything which has come before. A tightly scripted project might result in a fairly straightforward assembly but, in many cases, documentaries are created, or discovered, during post-production. The editing process provides opportunities to completely reimagine or reconstruct a film, to achieve new creative or intellectual visions not evident before.

Post-production is a period of practically unbridled creative and intellectual opportunities. Editing your footage together will, for better or for worse, show you the reality of your original vision. It will confirm your original genius or, particularly for first-time filmmakers, show the weaknesses and limitations present in your original plan. Like the unexpected setbacks that will have marked the production phase, this is nothing that cannot be overcome with some creative thinking and a willingness to reassess and rework your project.

Scripts can be rewritten in post-production. Shots not meant to go together can suddenly be used to create an entirely new or unexpected intellectual point. The rhythm of the finished film, which before post-production was only ever imagined, might turn out to be very different to that which you originally envisioned. In other words, you should expect your film to reveal itself to you throughout post-production — and you should expect the project to grow, change, and evolve.

Allow yourself to be responsive to your project's needs. By all means, focus upon achieving your original vision, if that continues to promise the best results, but be prepared to accept new possibilities in the editing bay.[4]

Editing occurs in roughly three phases: rough cut, fine cut, and finishing cut. The rough cut is the first version of the film that you will edit together and it should serve to give you a broad sense of what your finished film will look like, though it will likely have significant pacing issues, unfinished sequences, and a generally unpolished feel which will make it inappropriate to show outsiders. This is perfectly natural and you should not worry about producing a rough cut that does not yet feel like a film. The important thing is that you have a version of your film that you can assess and, with a little imagination, refine into a more satisfying state.

---

[4] An excellent introduction to the post-production mindset was written by Walter Murch, the editor of *Apocalypse Now* (1979). Whilst some of the technical information, even in the updated edition of Murch's book, is now out of date, the theory and ideology that he discusses certainly is not. See Walter Murch, *In the Blink of an Eye. Second Edition* (Los Angeles: Simlan-James, 2001).

That is not to say that disaster will not strike. Rough cuts are not always successful and may well demonstrate significant structural failings in your project, which you will need to address. If this occurs, know that you are in good company. The rough cut of George Lucas's *Star Wars* (1977) was a disaster — and its 2016 prequel, *Rogue One: A Star Wars Story* likewise required significant reshoots to reconstitute it into a form that pleased its studio and distributor.[5] Despite the setbacks, both of these films ultimately recovered and, at least in the case of *Star Wars*, resulted in a piece of era-defining cinema.

If a rough-cut of your film reveals serious issues, reassess and rebuild. Significant rewrites may be required and, possibly, the collection of new material (re-entering production, essentially), all of which might prove disheartening. If the result is an intellectually deep and effective film, however, it will be worth the additional effort.

Post-production can require bold decisions not envisioned during the pre-production or production stages. To that end, be prepared to edit around the material that works most effectively. Filmmakers should not be afraid of cutting material that does not add intellectual weight to the final project. Heart-breaking though it may be to remove a cherished sequence, it may be necessary for the good of the production. Filmmakers should thus be ruthless in the post-production process — ruthless with their emerging edit, with their pre-existing vision, and with the footage they have collected.

Once you have created a rough cut of your film with which you are broadly happy, you can begin working on your fine cut. At this stage in the process, you should pay particular attention to the timing of individual edits and the overall rhythm of your film. You should aim to ensure that your audience forgets that it is watching a film. Cuts should not draw attention to themselves and the audience should be engaged throughout. During this stage of the editing process, you should pay particular attention to the feel of your final film: does the audience receive all of the information they require at the right time and in the

---

5   *Empire of Dreams*. Directed by Kevin Burns and Edith Becker. Los Angeles: 20th Century Fox, 2004; and Aaron Couch. 'Tony Gilroy on "Rogue One" Reshoots: They were in "Terrible Trouble"', *The Hollywood Reporter*, 5 April 2018, https://www.hollywoodreporter.com/heat-vision/star-wars-rogue-one-writer-tony-gilroy-opens-up-reshoots-1100060

correct sequence; are there lulls wherein their interest may wander; could sequences be improved with sharper editing?[6]

The introduction of your final music choices and a well-developed soundscape should start to give your film a close-to-finished feel (see chapter twenty-three). Music should be present in both the rough and fine cuts, but in the latter stage it should be presented as it will ultimately appear in your final film. Depending upon the type of film you are creating, the music you use may well add significant depth to your work. If this is the case, it should be fully evident in the fine cut of your film.[7]

The final stage of the post-production process, the finishing cut, will see you adding the final polish which will complete your project. Any place-holders will all need to be removed and replaced with their final elements. Cuts will need to be finalised and any problematic moments or sequences will need be resolved or removed. Audio will need to be balanced and tweaked, to ensure that spoken-word sections are clear and audible; the music should complement your work, but it should not overwhelm it. The rough edges, in other words, should be removed in this final editing phase. The journey you commenced at the start of pre-production will now have reached its conclusion.

Your film will now be complete.

---

6   Ken Dancyger, *The Technique of Film and Video Editing: History, Theory, Practice* (New York and London: Focal Press, 2011), pp. 327–40.

7   Steve Saltzman, *Music Editing for Film and Television: The Art and the Process* (Burlington: Focal Press, 2015).

4. Concept and Planning

Fig. 5. Watch the second lesson in our documentary-making course. http://hdl.handle.net/20.500.12434/43f4c29c

There are many ways for you to approach documentary production. Some are ostentatious and difficult to achieve, whilst others will require little more than a camera, a microphone, and a small number of interview subjects. There is no standard model to follow, and the nature of the medium grants huge amounts of freedom. Much is achievable if you are willing to invest your time in achieving a particular vision.

That being said, there are four fundamental schemas you may wish to consider at the outset of your filmic endeavours. These models are not the limit of what scholarly films can be, but they are a solid foundation upon which you can begin to formulate your own project.

Schema One — Essay Films

Perhaps the most comfortable model for many scholars is one that closely emulates the type of written work with which they will likely be familiar. Essay films can be constructed around commentary tracks, which might include discussions or analysis similar to that found in traditional academic texts. Such films tend to include a visual element, or set of elements, which interact with the commentary track. This imagery can be abstract and symbolic, or it can be a more literal representation of the discussion at hand. In either case, essay films should not merely be an academic essay set to a visual montage. The visual elements should help to deepen the arguments and discussions at hand; they can be illustrative, serve as counterpoints, or offer an alternative intellectual discourse which interacts with the commentary track in stimulating and engaging ways.[1]

Perhaps the most famous example of an essay film is Orson Welles's *F is for Fake* (1973) but, as one might imagine from the director of *Citizen Kane* (1941), Welles's work achieves significant depth and is not easily emulated.[2] Instead, inexperienced filmmakers might be better served by considering Mark Cousins' *The Story of Film: An Odyssey* (2011). In this series it is Cousins' own commentary, working in tandem with the appropriately symbolic footage, which delivers the greater part of the analysis.[3] The result is an accessible and engaging piece, which demonstrates how a well-constructed script defies the need for complex set pieces. A more abstract example, principally thanks to its minimalist deployment of commentary, is Tony Silver's *Style Wars* (1983). Documenting the emergence of hip-hop and, in particular, graffiti culture in New York City, it is a wonderful example of how a filmmaker can use the world around them to create visually

---

1 Defining essay films, as has been done here, is problematic. There is significant discussion about the nature of essay films and the definition given here is certainly more restrictive than that used by other scholars. For a discussion on this, see Kevin B. Lee, 'Video-Essay: The Essay Film — Some Thoughts of Discontent', *Sight and Sound*, 22 May 2017, http://www.bfi.org.uk/news-opinion/sight-sound-magazine/features/deep-focus/video-essay-essay-film-some-thoughts. See also Elizabeth Papazian and Caroline Eades (eds), *The Essay Film: Dialogue, Politics, Utopia* (London and New York: Wildflower Press, 2016).
2 *F is for Fake*. Directed by Orson Welles. London: Eureka Entertainment, 1973.
3 *The Story of Film: An Odyssey*. Directed by Mark Cousins. Edinburgh: Hopscotch Films, 2011.

rich, in-depth discussions.⁴ Comparing and contrasting the first episode of *The Story of Film* with *Style Wars* should prove instructive for inexperienced filmmakers with strong ideas but limited resources.

Schema Two — Discussion/Interview Films

If the video essay is built around the filmmaker's thesis, the discussion/interview film differs in that it is instead built around a thesis (apparently) created by the film's key subjects. In many ways, *Style Wars* more comfortably fits into this category than that of the video essay, but its ability to appear in both highlights the fluid nature of the boundaries that separate these schemas.⁵ Rather than building a film around a written piece, the discussion/interview film instead places the emphasis upon verbal exchanges with third parties. In this model, interviewees appear to shape and guide the piece's thesis, though that is almost certainly not the case. The filmmaker-scholar's power, in this instance, comes from the questions they ask of their subjects, the context in which the interviews/discussions occur, and the way the resultant materials are assembled during the editing process. This model can accommodate a discussion with a single, particularly compelling subject, or it can contrast and compare ideas by juxtaposing dialogue.

Requiem for the American Dream (2015) is a film built almost entirely around a discussion with famed scholar and activist, Noam Chomsky. Whilst not always desirable, this model nonetheless demonstrates how an interview with a single individual can result in a deep intellectual inquiry — particularly when the film's intended audience is already very familiar with its principal subject.⁶ *The Fog of War* (2003) is likewise constructed around a single interview, with former U.S. Secretary of Defence Robert McNamara. Audio outtakes presented at the start of the film make it clear that McNamara had a very specific agenda, which he pursued throughout the project — a revelation that helps the audience to frame his later testimony.⁷ Both of these films show how discussions

---

4 *Style Wars*. Directed by Tony Silver. New York: Public Arts Films, 1983.
5 Ibid.
6 *Requiem for the American Dream*. Directed by Peter Hutchison, Kelly Nyks, and Jared P. Scott. El Segundo: Gravitas Ventures, 2015.
7 *The Fog of War: Eleven Lessons from the Life of Robert S. McNamara*. Directed by Errol Morris. Culver City: Sony Pictures Classic, 2003.

with single subjects can create exciting opportunities to capture discourse that is so compelling it can serve as the fulcrum around which the rest of a project can be constructed.

Schema Three — Full-Production Films

Essay films and discussion/interview films, at least as they have been described here, can be created with minimal resources. Full-production films, however, are a much more ambitious undertaking. Such a project would aim to mimic or innovate upon the larger-scale productions commonly consumed by broad audiences. These films can include a variety of complex visual elements, such as historical re-enactments, animations, dramatisations, and other elements created solely for the film project. Collaboration, to one degree or another, is likely to be required in order to achieve such cinematically ambitious ends — but by carefully planning a project, more ambitious set-pieces can be achieved.

There are innumerable examples of full-production documentaries to which scholars can look for inspiration. One particularly noteworthy example is the BBC's ostentatious *Wonders of the Solar System* (2010) series. Whilst we are not qualified to pass comment about its scientific worth, its use of music, computer-generated animation, and exotic locales provide a level of spectacle that suitably mirrors the series' epic scope.[8]

Schema Four — Subjective Explorations

Documentaries have the capacity to differ substantially from typical academic texts. Unlike a journal article, there is greater scope within a filmic framework to explore an author's subjective and personal relationship with their topic. Whilst academic writing can indeed be a place for personal reflection, films offer an opportunity to capture subjective moments as they occur. They also offer the opportunity to openly explore the author's subjective relationship with a situation. Academic writing may not be the ideal forum in which to reflect on one's emotional relationship with a topic — film, however, can provide a powerful vehicle to engage in such a discourse.

---

8 *Wonders of the Solar System*. London: BBC, 2010.

Orson Welles' *F is for Fake* is a masterclass in using film to explore one's subject from a range of perspectives, including personal and subjective positions. Welles rejects the authority of the author, in part, by making himself (and his implied authority) central to the audience's experience. Welles spins an elaborate tale about Pablo Picasso and his relationship with a dealer of forged artwork. The tale is, Welles eventually confesses, a forgery, but it was a lie told to reach a deeper truth. By making Welles an icon of authority, his ultimate confession carries all the more weight. The audience has, under Welles' direction, experienced the power of the fake. As a result, they are in a position more fully to appreciate the truths revealed by the art of forgery.

Whilst this is not a model that one should necessarily seek to imitate, there is much that can be learned from a close study of *F is for Fake*. Our own film, *Looking for Charlie*, breaks the fourth wall in a very different way, by drawing upon our own experiences with depression to make a deeper, albeit subjective, observation about our subjects — Charlie Chaplin, Buster Keaton, and the suicidal comedians who inspired them. This approach is certainly not for everyone, but the tools offered by the filmic medium are powerful, and they can be used in a multitude of unexpected ways.

Of course, there are schemas beyond those covered here. This is a foundation upon which you can build, not the limit of what you can produce. Indeed, the barriers that separate each of these schemas are fluid and likely to be contested — where one ends and another begins is a matter of subjectivity and taste. There is nothing to stop a filmmaker-scholar from creating an essay film that includes full-production elements or substantial discursive sections. These models are merely suggestive frameworks.

Achievability

We cannot define your project for you. Only you can conceive of the type of film you might bring into being, its intricacies and intellectual potential. If you are reading this book then, in all likelihood, you already have some type of vision for a scholarly film — a subject area, thesis, chronology, list of topics, and so on. That part of the process is entirely yours and, as such, this guide can offer little specific advice.

Still, it is worthwhile thinking through how the careful planning of your film can allow you to achieve your intellectual or creative goals: as you are conceptualising your film, you should always aim to keep at least one eye on practical considerations. By all means, explore the ways in which intellectual ideas can be visualised but do not forget that, sooner or later, it will be up to you to realise your vision. Do not curb your enthusiasm (or ambition), but work to ensure that your vision is an achievable one.

One way to ensure achievability is to think about the following three goals for your production — and then picking *only two* of them: quality, speed, and affordability.

The idea here is simple — in all likelihood you cannot make a film that is cheap, quick to produce, and of a high quality. You can, however, produce a high-quality film on a small budget; but it will likely require a lot of time. Likewise, this model tells us that you can make a good film in a short space of time; but doing so will not be cheap. Perhaps most importantly, it suggests that you can create a cheap film in a short space of time; but it will likely be of poor quality. In order to ensure quality, a significant amount of money or a significant amount of time will need to be invested.

Hardly scientific, this model is, at best, advisory — but it makes a good point. If money is not an object, there is little that you cannot achieve by hiring the correct equipment and crew. Assuming, however, that you do not have a substantial budget (perhaps you do not have one at all), the option of buying your way out of a problem will not be available to you. That being the case, you need to accept that time, rather than money, will be your principal currency; time to learn how to use your equipment; time to grow your skills as a writer, editor, interviewer; time to allow you to work with the goodwill of those people you invite to be a part of your production; enough time to ensure you will not have to make undue sacrifices in your personal or professional life.

Time does not entirely negate the need for money, but it can certainly help. Some things will simply not be achievable on tiny or non-existent budgets — but some version of your vision may be, if you are willing to take the time required to think around the problems at hand. An excellent case in point is Peter Watkins' docudrama *Culloden* (1964), which sought to re-create, and then document, the Battle of

Culloden from 1746. Thanks to careful planning and imaginative use of camera angles, Watkins was able to give the impression that his film was shot amid an unfolding battle — despite him only having access to a small number of amateur actors and extremely limited resources. By focusing attention on individual moments within the battle and never attempting to depict its full scale, Watkins was able to take audiences into the unfolding conflict, speaking to important subjects and exploring their perspectives as events appeared to unfold in real time around them.[9] The result of this subtle subterfuge really is quite remarkable and effective.

On first blush, creating a documentary about a battle that involved 15,000 people might seem like the type of enterprise that would require a massive budget. Indeed, it might even appear an impossible task for most independent filmmakers. But by carefully utilising the available resources, Watkins demonstrates that it is possible to carry out such a challenging brief. *Culloden* is far less interested in depicting the mechanics of the Jacobites' defeat or the scale of the battle than it is with exploring the attitudes of those involved in it. As a result, much of the film is built around faux interviews with important leaders and lower-level participants in the battle. Military manoeuvres are depicted, but such scenes focus upon small groups, representative of the larger whole. These moments are then intercut with on-the-ground 'interviews' whilst, in the background, action (which could be accomplished with only a few extras) carries on.[10]

The intellectual drama of *Culloden* comes not from the thrill of seeing an extensive battle depicted by an army of actors; it comes from the contrast between ordinary soldiers and their leaders, particularly on the Jacobite side. Structural inadequacies in the organisation of the Jacobite forces are brought to the fore, the arrogance of their leadership is demonstrated, and, as a result, the ordinary solder is cast as a type of tragic figure. Whether one agrees with it or not, the film has a clear thesis which it makes with force. The scale and scope of the battle did not need to be depicted because it was in intimate moments that the film's case

---

9 *Culloden*. Directed by Peter Watkins. London: BFI, 1964
10 For a personal reflection on *Culloden*, see Alex Cox, 'Not in Our Name', *The Guardian*, 9 July 2005, https://www.theguardian.com/books/2005/jul/09/featuresreviews.guardianreview12

was made. One is free to disagree with the film's argument, or its use of fictive evidence (staged interviews), but denying its effectiveness would be much more difficult.

The battle became a background detail in *Culloden*, which, in turn, allowed the overall film to feel much larger than its component parts. As a result, it is a study of what can be accomplished when available resources are utilised carefully and imaginatively. If you wish to recreate a historic episode, do so; but like Watkins, use your available resources with care. Construct a film that utilises (rather than suffers as a result of) these limitations. Use local locations, students of drama and theatre, amateur actors, readily available costumes, the cameras at your disposal, and so on. If you wish to emulate the *Culloden* style, a camera capable of shooting in a shallow focus (allowing background action to be blurred) will make it easier for you to create the illusion of background movement without requiring highly detailed costumes or props for your background actors. This would also allow the same actors to be employed in numerous roles as their faces, bodies, and costumes will be so blurred that they will be functionally unrecognisable. An easily reached location may not be ideal if it is not the spot that is supposed to be depicted, but a shallow focus can be used to eradicate unwanted details that might otherwise identify the setting. In such a way, a relatively small number of actors could, in a carefully planned shoot, be used to create an illusion far grander than initially seemed possible.

Case Study — Signals

In 2017, we began work on a short documentary about the maritime history of the Scottish town of Arbroath (working title — *Signals: Scotland and the North Sea*). The opening sequence of the film depicted the arrival of a group of eighteenth-century smugglers. In the most ambitious version of this scene, a small rowing boat would have landed on a secluded beach in the dead of night; a smuggler crew would then have begun unloading their wares, before dragging various chests and barrels up the steep path from the beach to some nearby clifftops. Crude wicker torches would have lit the haggard and sea-worn faces of the crew; the light dramatic, the atmosphere oppressive.

4. Concept and Planning 49

Fig. 6. Our smuggler crew prepare to ascend the Seaton Cliffs in Arbroath.

Fig. 7. The scenery around the town of Arbroath is inherently dramatic, adding significant production value to any scene shot there. No tall ships were required to give this scene a sense of drama.

Unfortunately, such a dramatization would have required a substantial budget: the cost of lighting a scene at night, safety marshals to ensure the wellbeing of cast and crew in low-light and low-temperature conditions, a support vessel with trained lifeguards, a wide variety of props, and so on. As originally envisioned, the scene was simply not achievable within our available budget, but that did not mean that the essence of this scene could not be realised.

The first challenge we faced was populating the scene. We reached out to local amateur dramatic societies and recruited three actors to play our crew of smugglers. As an arrival by boat would have been cost-prohibitive, we instead envisioned a much simpler solution: the camera, close to the ground, water lapping against the sand. Feet, clad in old boots, step into the shot. They shuffle through the scene, the legs and feet of our crew struggling as they drag their wares through the frame. Finally, we cut to a more traditional waist-up perspective. Not as dramatic as an arrival by boat, but vastly cheaper (and just as effective).

Rather than insist on the inclusion of elements that were either costly or difficult to execute, we instead decided to work with the resources that were freely available and easily accessible. We had ready access to a stretch of coastline, consisting of cliffs, beaches, coves, and caves. For no outright cost, we were able to film in a location filled with texture and inherent drama. Our actors' outfits were provided by a local theatrical costuming business and a large chest was purchased to serve as the scene's main prop. A friend of the production, with experience in the theatre, volunteered their services as a makeup artist. What could have been an expensive and difficult scene ended up costing very little. Significant effort and goodwill was required to realise it, but the final sequence captured the substance of the original vision.

The shoot was efficient and effective. We had already storyboarded the entire sequence, generating a list of shots that we needed to capture on location. We utilised Google Maps and other such resources to map out precise shooting locations, calculating factors such as travel time, rest time, and so on. We also consulted weather and tidal reports to ensure a safe and comfortable environment for the cast and crew. We started shooting at 9am, ensuring sufficient natural light. By following the production schedule that we had created in the weeks prior to the shoot, we were able to shoot efficiently — and in the knowledge that we would capture all of the coverage (necessary shots) that we would require in the editing process.

Planning

Complex sequences should be pre-planned and, where necessary, rehearsed. The actual shoot should be the culmination of a process that has been thoroughly planned. Do this effectively and you will be able to extract every ounce of value from the time, and resources, you have available to you.

Storyboard pre-planned sequences. Combine photographs with simple renderings of your characters or subjects to create a visual guide to all of the shots you will need to capture. Storyboarding may well intimidate those of us who cannot draw effectively. This need not be the case, however. Take still photographs with stand-ins, either on location or at home, to create a series of still images for your storyboard, or utilise one of a number of apps that allow you to use stock art (including 3D models) to create storyboards. Examples of these include Previs Pro and Shot Designer for smartphones and tablets. This will allow you to pre-plan all of the different shots you will need once you are on location. From your storyboard, generate a shot list. Organise this list into an efficient and achievable shooting schedule.

The creation of a sequenced shot list will thus generate a schedule of actions, a clear plan that will lead you to capture all of the necessary raw footage you require. You must now study this plan and calculate the time necessary to execute each individual action or shot.

Remember, cameras must set up in the correct locations, shots must be composed, and settings adjusted; the featured actor must be wired for sound (if they have dialogue); the audio equipment be set to record; extras must be directed; discussion between the director and their cast and/or crew may follow. Once the camera starts filming, it will take a period of time to achieve the precise shot or performance you desire — perhaps it will take several attempts. Once the shot is completed, this entire sequence of events will need to be repeated. Many different shots from many different angles may be required to create a usable bank of footage. On some occasions, the scene will be filmed in close-up. At other times the camera will be further away from the action. In each instance, cameras will probably need to be moved, lighting adjusted, new direction given to a performer, and so on. In practical terms, this means that you will need to move and re-frame your cameras and actors multiple times. There is a time implication for each new setup.

In all, then, a sequence designed to take up no more than a few minutes of screen time might easily take three to five hours to shoot, or even longer — perhaps significantly longer. Even if you are able to move and setup your equipment with military-grade efficiency, actors will give uneven performances and lines will be forgotten. Tempers will become frayed as the cast and crew grow increasingly tired. They may become fatigued and require rest. The lighting, particularly if it is natural, might change in unexpected ways. Many factors can lead to a seemingly simple sequence becoming a rather drawn-out or difficult affair.

But there are economies of scale at play that can help you to optimise your time. If you have a camera setup that you intend to use for several shots, shoot all of those sections together, regardless of whether this is consistent with the internal chronology of your scene: film shot 1 from Sequence A, shot 3 from Sequence B, and shot 2 from Sequence C, and so on. You should plan this ahead of time using your shot list, which, at the very least, should attempt to anticipate how much time each camera setup and performance will take. Early in your filmmaking career, you will certainly underestimate the time required. Indeed, by working through the practicalities of the process and creating your shot list (with anticipated times) you may discover that you simply cannot shoot all of your desired footage in the available time. If this occurs, a change of approach will be required. But at the planning stage, this realisation is unlikely to upend your production. If this realisation occurs on location, however, where your ability to adapt may be more constrained, more significant problems may follow.

With a detailed shot list and schedule, you will now be in a position to compile a list of the precise resources that you will require to complete your sequence. Compile a list of every piece of required equipment, taking care to ensure that you include necessary accessories, such as tripods or a variety of different lenses (see chapter seven). You will also need to generate a list of collaborators: does your planned sequence require you to hire or work with a large number of other people? If so, it may be necessary for you to reconsider your sequence; a large number of participants will increase the complexity of a shoot, and likely slow it down significantly. The more complex the machinery, the more prone it will be to breaking down.

With a resource list, you will be better positioned to recognise if your planned sequence remains achievable and within your means. Discovering that a relatively simple sequence might require a large number of actors who, in turn, would all require costumes, props, and food, may well require you to rethink your plan. In such a case, once again reflect upon the intellectual or dramatic essence of your sequence and consider how it can be achieved with the resources that are within your means. Remove superfluous action or difficult-to-achieve shots. Consider alternatives that are easier and quicker to produce, but are just as effective and intellectually satisfying.

5. Collaboration

Making a documentary is an immersive experience. You are creating a truth into which you put your heart and soul. It can be lonely process, but it can also be a shared experience. In an increasingly digitally-driven world where filmmaking technologies are democratised, more affordable, and increasingly user friendly, and in a technological environment in which connectivity is the norm, working collaboratively is easier than ever before. Pop songs with multiple voices are produced without the artists meeting in the studio; individual recordings are made in smaller studios — often at home — and amalgamated on a computer somewhere else entirely. This is twenty-first-century media production. In the academy, such collaborative digital processes promise exciting new intellectual opportunities.

Working collaboratively is a wonderful thing. It provides multiple ideas, perspectives, visions, and skillsets, which can be explored using the specific grammatical opportunities offered by digital filmmaking. Working on a media project with a friend or colleague requires an additional level of planning, however. Your documentary-making collaborator may share a vision with you, but it is unlikely to be identical to your own. Before you pack up your equipment and head out on location you need to discuss, in an open and frank way, what it is that you are trying to achieve. This may seem like an obvious step in the process, but it is too easy to assume that you already share a cohesive vision when there are, in fact, problematic differences between what you and your partner(s) hope to achieve, and how you plan to achieve it. In a process as complicated and involved as documentary production, such divergent ideas can cause significant problems if they are not resolved in advance.

This applies not only to the overall vision for the film, but also to the finer details, such as the type of shots you need to capture. To that end, ask yourself the following questions:

- Where are you going?
- Do you need a crew; if so, how large will it be?
- How will creative and intellectual responsibilities for the project be divided?
- What mechanisms are in place to manage disagreements?

When we began working on *Looking for Charlie*, we had to develop clear roles which served the project. We both co-directed and co-wrote the film, but Darren was to serve as editor and Brett as the film's producer; roles that played to each of our strengths. We also had to ensure we were on the same intellectual page. To facilitate this, we exchanged reading lists and set aside time to discuss the literature surrounding our chosen topic (life in the silent era), working through our individual thoughts and developing a shared direction for the film. The work was based largely on research Darren had already carried out — but Brett offered new ideas and perspectives that would shape how the film would ultimately evolve and develop.

Fig. 8. Watch the trailer for *Looking for Charlie*. Scan the QR code or visit http://hdl.handle.net/20.500.12434/2313fcf2

In particular, our collaboration allowed us to explore more subjective, personal aspects of the film's core themes — depression and recovery. That would have been difficult for either of us to recognise or pursue as individuals, not least because we found the filmmaking process to be a type of catharsis during a very challenging period in both our lives. As friends, we were able to support each other; as collaborators, we were able to recognise how our own personal experiences reflected key themes in the film. The parallels between the film's subject matter and our own experiences created new discourses between us, some of which ultimately informed or appeared in the final film. What could have been a relatively simple documentary about life in the silent era, instead became a much more personal reflection on surviving depression as seen through an early filmic lens.

When you are choosing a subject for your film, choose something that is close to you, a part of you, and do not be scared to open yourself up to your audience — or to your collaborators. Professor Green's film about depression and suicide, *Suicide and Me* (2015) was made much more interesting and engaging thanks to his personal story about the loss of his father to suicide and his subsequent struggles with

Fig. 9. Shooting on location at Cirencester, behind the scenes at Gifford's Circus for *Looking for Charlie*. L-R, Darren R. Reid, Brett Sanders, and our subject for the day, Tweedy, a professional clown.

depression and search to understand his father's actions.[1] Do not be afraid of subjectivity; we cannot always detach ourselves from the issue we are documenting and an audio-visual grammar offers opportunities to explore such subjects beyond the framework typically provided by academic papers. Choosing one's collaborators should thus be done with care. Filmmaking can be a very personal and challenging process. Plot, plan, and communicate.

Unlike *Looking for Charlie*, *Aftermath: A Portrait of a Nation Divided* (2016) started life with a clear sense of objective purpose. We would not indulge our own subjectivity. Instead, we sought to take the pulse of New York during a charged and contentious electoral cycle, soliciting the subjective views of ordinary Americans in a dispassionate and honest way. To achieve this, we worked to ensure that we had a clear, shared vision — rather than having our own story to tell, we would allow our subjects to lead the narrative. We were to be responsive to the story that New York wanted to tell.[2]

Regardless of what kind of film you intend to make, it is crucial that you organise yourself and your collaborators effectively. You will only have a limited amount of time in the field; you are limited by the battery life of your cameras, and by other factors such as light. If you have a large number of collaborators (a crew), organise them into smaller units with specific tasks. One team might be tasked with finding locations, another with shooting interstitial material, and so on. When making *Aftermath* we divided ourselves into units, which allowed us to run parallel tasks, maximising our time in the city. One team was responsible for filming our interviews, another looked after our interviewees, and another captured shots of the city. What would have taken a single unit three days could thus be accomplished in less than half that time.

This is where working collaboratively offers great advantages. Having a wider toolkit of skills, and personality types, is a key advantage to working as part of a team. On both *Looking for Charlie* and *Aftermath*, we, as co-directors and project leads, each brought skills and knowledge to the project, but we also had our crew's skills, knowledge,

---

1   *Professor Green: Suicide and Me*. Digital Stream. Directed by Adam Jessel. London: BBC, 2015.
2   *Aftermath: A Portrait of a Nation Divided*. Digital Stream. Directed by Brett Sanders and Darren R. Reid. Coventry: Red Something Media, 2016.

and enthusiasm to draw upon. Identifying the skills that you and those around you possess is really important. We recognised immediately which of us possessed a passion for design and which of us possessed an eye for detail. We knew what we wanted to achieve, shared a vision, and understood our individual strengths and weaknesses. As a result, we were able to work together in a complementary way. We shared writing and directorial responsibilities, but Darren served as editor, and Brett as lead producer.

Trust is the natural product of close and effective collaboration. When Darren made *Keepers of the Forest: A Tribe of the Rainforests of Brazil* (2019), Brett was an important part of that project's post-production process.[3] The film had been made when an unexpected (and time-sensitive) opportunity presented itself, thus preventing full horizontal collaboration. Post-production, however, presented the opportunity for broader collaboration, with Brett ultimately serving as the film's executive producer and creative consultant. Modes of collaboration may vary, but effective partnerships should be maintained, nurtured, and utilised wherever possible.

Filmmaking creates opportunities to work with a wide range of potential collaborators, not just those who are responsible for the overall creative and intellectual integrity of a project. Every camera person, production assistant, or sound recordist is a collaborator, even if their contribution is focused and specialised. When making *Aftermath*, we combined the production process with a learning experience; our crew was comprised of undergraduate history students who were looking to broaden their CVs. We recognised that two of our crew possessed specific talents: one had an excellent eye for detail and for the framing of shots; the other had excellent people skills, as well as a good technical understanding of the camera equipment. As a result they were each given the role of Assistant Director, and throughout that project each was delegated tasks that best reflected their abilities. As we filmed interviews in Harlem and Wall Street, for instance, we were able to dispatch one unit, under the direction of the relevant Assistant Director, to find interesting shots that we could use to lead our audience through our portrait of New York.

---

3   *Keepers of the Forest: A Tribe of the Rainforests of Brazil*. Directed by Darren R. Reid. Coventry: Studio Académé, 2019.

We likewise invited the rest of our students to think about the remaining roles available and where their own skillsets lay — interviewing, fixing, sound recording, and filming. This allowed us to place people in the most appropriate roles, harnessing organic enthusiasm and pre-existing skillsets. Self-confident members of the crew approached New Yorkers, asking them if they wanted to take part in our project, whilst others interviewed them, recorded sound, operated cameras, and so on. As the shoot progressed, we provided opportunities for crew members to experience different roles before settling into positions that reflected their core strengths. As a confidence-building exercise, this helped to reinforce their strengths.

Our crew ultimately settled into the following structure:

- Co-Directors x 2 (Brett and Darren).
- Assistant Directors x 2.
- Fixers x2 [Members of the crew responsible for carrying out whatever minor tasks are required by the directors].
- Interviewers x 4.
- Camera operators x 8.

Once we had wrapped up the shoot and returned home, we were able to start the process of assembling our footage. In all of our projects we spent hours watching raw footage, a tedious but essential part of the filmmaking process. Clear your diary of a day, or days, buy junk food, and prepare to settle in. For every hour of footage we produced and watched, we used perhaps 10% of it in the final cut. Whilst the end product will look polished, professional, and glamorous, the process is often less so. Trawling through your footage is the least stimulating part of the process — planning is fun and imaginative, as is story-boarding. Filming on location is also an amazing experience. Not so trawling through hours of interstitial material, looking for *that* five seconds of footage. Still, with a collaborator the process was somewhat more creative than it otherwise might have been; an informative intellectual discourse can emerge even during tedious tasks.

Remember to organise your recorded material well. Failing to do that will make this part of the process incredibly difficult.[4] Having said

---

4   Barry Hampe, *Making Documentary Films and Reality Videos* (New York: Henry Holt and Company, 1997), pp. 279–83.

that, this is also the part of the process where some element of your production's truth is realised. In *Looking for Charlie* we were using film to revive the memories of two largely forgotten comedians. We gave them a voice, and highlighted their significance to the world that had forgotten them. In *Aftermath* we gave a voice to New Yorkers who were, at that time, trying to understand what it meant to be an American in the era of Donald Trump. In our current project, *Signals: Scotland and the North Sea*, we are only just discovering the truths held by our material. Watch your material together; just as you plan and execute the capture of your footage together. Make every part of the process a collaborative exchange and you will create a framework in which you will consistently discover (and build upon) new ideas.

Working collaboratively is an exciting proposition — you share skills, adventures, and tasks. Our filmography is the result of our love of collaboration. We would not have captured as much footage, or as many interviews, if we had worked independently. Nor would we, particularly with *Looking for Charlie*, have been able to realise a project that became so large. It consumed more than three years of our lives, shooting in half a dozen major locations spread across three continents. Mutual support kept us going at times when, as individuals, we almost certainly would have given up or settled on something far less ambitious.

Collaboration, then, can help you to create intellectual and narrative studies of far greater scope than you might otherwise be able to accomplish on your own.

6. Precedent

Just as with traditional humanist writing, documentaries are created within a methodological context. Filmmaker-scholars will continue to draw upon the research and literature of their peers, rooting their works in a deep understanding of the scholarship on a given topic. But they must also work self-consciously within the framework created by the medium they hope to utilise. Just as scholarly literature will frame and inform your ideas, so too should filmic precedent inform the look, feel, and communicative tools drawn upon by the filmmaker-scholar.

Watching a wide range of films, both drama and documentary, will provide you with many different models that can be emulated, contested, or subverted. Whilst no single viewing list can cater to every taste or permutation of intellectual desire, we have found that the following films have proven particularly provocative, insightful, and inspiring: *F is for Fake* (1975) by Orson Welles, *The Story of Film* (2011) by Mark Cousins, *Confessions of a Superhero* (2007) by Matt Ogens, *Style Wars* (1983) by Tony Silver, *Best Worst Movie* (2009) by Michael Stephenson, *Capitalism: A Love Story* (2009) by Michael Moore, and *Exit Through the Gift Shop* (2010) by Banksy. You may draw inspiration from other sources. Indeed, we thoroughly encourage this. It does not matter if you are inspired by the same material as ourselves. What matters is that you build a sense of what the medium is capable of and what you can contribute to it. This chapter is merely a starting point in that process.

Both fiction or non-fiction will expose you to a wide range of visual grammars, dialects, and techniques. Every film is an essay on the many ways to succeed or fail at communicating ideas via an audio-visual medium. The controversial dramatic series *24* (2001–2010) was shot in a quasi-documentary style, to underline the sense of reality it sought to foster, but there is nothing to stop documentaries from, in turn, borrowing from it. With its problematic look at terrorism and anti-terrorism, the series might not be an obvious inspiration for a scholarly film, but its split-screen simultaneous depiction of parallel events allows for the complexity of individual moments to be explored in detail.[1] To draw inspiration from *24* — or any drama — is to recognise an effective audio-visual grammar, one that can create a specific impression upon an audience and might add value to an on-screen intellectual discourse when it is appropriately retooled. It does not imply an acceptance of the ideology behind that original project. Whatever films or sequences inspire you, attempt to innovate or build upon the techniques you see, using them in new contexts or in different ways. You should not aim to replicate what has come before, but you should be prepared to respond to it.

In his 2007 film, *Confessions of a Superhero*, Matthew Ogens cuts from meticulously photographed interviews with his main subjects (struggling actors who play superheroes on the Hollywood Walk of Fame) to on-the-ground documentary footage of their everyday lives. This allows for more traditional documentary segments to be framed by deeper, more reflective insights, the unconscious (the happening) versus the conscious (the reflection on the happening). The approach resembles, in an abstract way at least, that of Woody Allen; the dichotomy between Allen (the character) and Allen (the narrator). That is not to say that *Confessions of a Superhero* resembles any particular Allen film — it does not.[2] But the interview segments of *Confessions of a Superhero* nonetheless serve a similar function as, say, Allen's frank voice-over, in *Annie Hall* (1977): the happening versus the reflection; the moment versus hindsight. Drama should not necessarily be imitated by filmmaker-scholars, but that does not mean that moments or devices used within dramatic films cannot inspire them.

With *24*, drama borrowed from documentary for the sake of style. With *Confessions of a Superhero*, documentary borrowed from drama for the sake of substance. From a functional perspective, then, there is

---

1   There is much to be said about the problematic politics of *24*, but for a very brief insight see Jane Mayer, 'Whatever it Takes: The Politics of the Man Behind "24"', *The New Yorker*, 19 February 2007, https://www.newyorker.com/magazine/2007/02/19/whatever-it-takes and Gazelle Emami, '24 is Back to Make you Fear Muslim Terrorists Again', *Vulture*, 2 February 2017, http://www.vulture.com/2017/02/24-legacy-muslim-terrorists-terrible-timing.html

2   *Confessions of a Superhero*. Directed by Matthew Ogens. Toronto: Cinema Vault, 2007.

no hard or fast line between documentary and non-documentary and, as such, each piece of media consumed by the filmmaker is one that is potentially filled with important lessons. The opening sequence of *Manhattan* (1979), the parallel action of *24*, the carefully shot interviews in *Confessions of a Superhero*, all are valid precedents.[3]

Quite naturally, the works of other documentarians should provide a particularly rich source of inspiration and counterpoint, particularly as they relate to how you can use and assemble your footage. Ken Burns's monumental series *The Civil War* (1990) is, its intellectual content aside, a masterful demonstration of elegant simplicity. The commentary, which leans from ostensibly neutral to openly sentimental, is typically delivered over a series of still photographs. Cameras pan or zoom, in a slow, gradual sweeps, revealing new details in these still images, in much the same way that a camera panning across live action might. The change of the voice, from that of the narrator to an actor reading a historical source (in character) adds to the overall atmosphere. No expensive historical re-enactments were needed to stir an emotional response in the series' audience. But as effective as the technique was, it has also become clichéd. It is so characteristic of Burns's output that to imitate it would be to invite comparisons and accusations that, like Burns, you are romanticising, rather than analysing, your subject.[4]

Less sentimental, but no less manipulative, is 2007's *King of Kong*, from director Seth Gordon. It chronicles the tale of two duelling videogamers as they compete against each other (and themselves) to become the holder of the world record in a classic arcade computer game. The film principally revolves around the rivalry between long-time 'Donkey Kong' champion Billy Mitchell and challenger to the title, Steve Wiebe. In the film, Mitchell comes across as arrogant, cold, and more than a little bullish, the perfect villain to Wiebe's struggling, humble underdog. If *King of Kong* succeeds at anything, it is in the presentation of a tight,

---

3   Every DVD director's commentary is a documentary about how a film has been assembled, about the numerous decisions and hardships that went into the making of a given production. The making of a drama may not feel instinctively appropriate to the documentarian, but many of the decision-making processes faced by the creators of drama are faced by the creators of documentaries. Both use a similar set of methodologies and both seek to move their audience in some way.

4   For a range of academic responses to Burns's *The Civil War* see Robert Brent Toplin (ed.), *Ken Burns's The Civil War: Historians Respond* (New York and Oxford: Oxford University Press, 1996).

compelling narrative rooted in the excitement of the mundane and the universality of an underdog story. On the surface, at least, it is a powerful example of how deeply documentaries can entertain when they happen upon a set of compelling circumstances or subjects.[5]

King of Kong is immensely entertaining but, according to post-release interviews, some of the events depicted in the film did not occur as they appeared in the final edit. Throughout the film, it is constantly implied that Wiebe is struggling to overcome not only Mitchell's high score but his influence in the world of competitive video-gaming. The audience is led to believe that Mitchell's long-time record was being unfairly protected by the scene's vested interests when, in reality, Wiebe's record was accepted at a fairly early point in the process. The footage used in the film was carefully edited together, turning the real into a semi-fictitious reordering of evens, creating an impression so compelling that its audience would have little reason to doubt its veracity. That Gordon created his finished film from more than three hundred hours of footage is indicative of the many potential forms it could have taken. *King of King* tells a masterful story, but it is perhaps more important as an example of how far the medium can detach its audience from reality, even as the audience believes that the opposite is occurring.[6]

To be fair to Gordon, the creation of a fiction from reality is nothing new in documentaries. Robert J. Flaherty's landmark film, *Nanook of the North* (1922) claimed to give its audiences insight into the lives of an Inuk man and his family but, in reality, much of the material that appears on screen is staged or distorted. The result was a type of dramatisation of real life, a semi-mythical reimagining of the Inuit in the early twentieth century that was anachronistic and romanticised. It fed into larger racial-social images that celebrated pre-modern, but not modernised, indigenous peoples.[7] That *Nanook of the North* is clearly

---

5   *King of Kong: A Fistful of Quarters*. Directed by Seth Gordon. New York: Picturehouse, 2007.
6   "The Kings of Kong", *Retro Gamer Annual* 4 (2017), 47–53; Walter, 'King of Kong — Official Statement', *Twin Galaxies Forum*, 26 September 2007–2012 March 2009, https://www.twingalaxies.com/forumdisplay.php/406-The-King-of-Kong-Official-Statement?sort=dateline&order=asc
7   For an example of how *Nanook of the North*'s illusion of authenticity has worked, see Barbara C. Karcher, 'Nanook of the North', *Teaching Sociology* 17 (1989), 268–69; for a more critical discussion about *Nanook of the North* and the ways in which its representation of its subject people is problematic, see Shari M. Huhndorf, 'Nanook

sympathetic towards its subjects does little to dispel how problematic its core worldview is.[8] Emotional identification with its subjects was achieved, but only at reality's expense.

Fig. 10. *Nanook of the North* (1922), directed by Robert J. Flaherty.

Documentaries have much to learn from each other; lessons in how to achieve, and how to fail at, their respective tasks. That *Nanook of the North* can be talked about next to *King of Kong* speaks to thematic or methodological consistencies in the genre, if not in every individual documentary, from which you can draw lessons. Inspiration should not always be literal; one should not aspire to distort the truth in order to create a more compelling narrative, despite the long roots of that tradition. That some filmmakers have placed secondary importance upon creating a reasonable interpretation (and representation) of the truth should be a point of contention and reaction; the filmmaker-scholar should work against such approaches, not embrace or encourage them. In his 2003 acceptance speech for the Academy Award for Best Documentary, Michael Moore famously declared that 'we live in fictitious times'.[9] Though he was referring to the logic behind the

---

      and his Contemporaries: Imagining Eskimo Culture, 1897–1922', *Critical Inquiry* 27 (2000), 122–48.
8   *Nanook of the North*. Directed by Robert J. Flaherty. New York: Pathé Exchange, 1922.
9   Michael Moore, 'Academy Award Acceptance' (speech, Los Angeles, 23 March 2003).

forthcoming American-led invasion of Iraq, he might have just as easily been describing the state of the documentary genre. Taken as such, it is a comment worthy of much reflection.

At least with drama, there is (typically) no confusion about the fictitious nature of the events depicted on screen. The audience understands that they are watching a piece of drama and the events being depicted are a fiction that exists solely within the confines on the screen's frame. Camera movements (a slow zoom towards a face, turning a mid-shot into a close-up) in drama are openly, if not always obviously, attempting to elicit an emotional response from the audience, and the audience is, on some level, aware of this.[10] In documentaries, however, that is not always obvious, particularly as the viewer runs the risk of being swept up by powerful analysis and emotive imagery, which make a claim to objectivity and veracity. Techniques differ between fiction and non-fiction, but the results are often the same. Much can be borrowed from drama to create deeper, more engaging intellectual experiences; much can be discarded from documentaries to create a deeper, more meaningful candidate for the truth.[11]

The camera captures what occurs in front of it, but it is the filmmaker who constructs a film's truth, be it in a fictitious, hyper-real fantasy like *Star Wars* or in a documentary film like Michael Moore's *Fahrenheit 9/11* (2004). The ostensible goal of most documentaries is the attainment of objectivity, a dispassionate analysis of events that accounts for their causes and/or consequences.[12] In reality, whatever the tone a documentary takes, it is always deeply editorialised. Ken Burns's *The Civil War* is at least open in its sentimentality, even if the audience is not given the intellectual tools (in the series itself) to compensate for and deal with that in-built authorial bias. *King of Kong*, however, is significantly less open about the way in which it is manipulating its audience. In both of these cases, there is much filmmaker-scholars can learn by studying, if not imitating, these two examples.

---

10 Anthony J. Ferri, *Willing Suspension of Disbelief: Poetic Faith in Film* (Lanham: Lexington Books, 2007).
11 Perhaps the one clear exception to this is the historical drama, which is often viewed as containing some essential element of truth by a significant proportion of its audience. See Thomas Doherty, 'Film and History, Foxes and Hedgehogs', *OAH Magazine of History* 16 (2002), 13–15.
12 *Fahrenheit 9/11*. Directed by Michael Moore. Santa Monica: Lionsgate, 2004.

A more subtle approach to editorialising, though one that is no less dangerous, is taken by documentaries that utilise a neutral, observational tone. Tony Silver's 1983 film *Style Wars*, about the emergence of Hip-Hop culture, features only a tiny amount of commentary. Unlike films such as *King of Kong* or *Confessions of a Superhero*, there is no attempt made at constructing a character arc out of any of the people who appear in this film, giving *Style Wars* a contrivance-free feel. There remains, however, significant editorialising and authorial bias within the film. Whilst Detective Bernie Jacobs, who struggles against the proliferation of graffiti in New York City, is hardly a villain, he does represent the normative counterpoint around which the film is constructed. Unlike most of the film's participants, he wears a shirt and tie and, like the mainstream culture that the film aims to chide, he sees graffiti tagging (the focus of the film) as a nuisance and as an act of criminality.[13] As this is a film about tagging, Jacobs is implicitly criticised throughout — not wrong, per se, but limited in his vision because he, like most of *Style Wars'* audience, was ignorant of the social significance of the tagging movement.[14] Graffiti tagging might be illegal, but that does not, the film argues, make its adherents immoral.

Despite its neutral tone, minimal commentary, and its apparent ambivalence towards its subject, *Style Wars* has a clear message: graffiti tagging and wider Hip-Hop culture, cannot be judged by a binary right-or-wrong standard. It is a symptom of change and societal unease, not the cause; like all art, the film seems to say, tagging is about generating necessary social discourses which otherwise might go unheeded. All of this goes unsaid in the film, but is nonetheless communicated, in toto, over the course of its duration, a thesis delivered through atmosphere and immersion rather than words or explicit argument. *Style Wars* is a wildly effective and fascinating piece.

The film's use of contextual footage as a means of developing and communicating this discourse is inspired. Without ever saying so directly, Silver depicts New York as a type of ever-changing art gallery in which the struggles of the city's voiceless denizens are now able to find expression. Every subway car becomes a moving wall in this

---

13   Sharon R. Sherman, 'Bombing, Breakin', and Getting Down: The Folk and Popular Culture of Hip-Hop', *Western Folklore* 43 (1984), 287–93.
14   David Craven, 'Style Wars: David Craven in Conversation with...', *Circa* 21 (1986), 12–14.

living gallery, documenting gang rivalries, love affairs, and individual aspiration. For their part, the city authorities have a role to play in the evolution of this living artistic space, continuously struggling to wipe away all signs of the culture that *Style Wars* was so determined to expose. If *Style Wars* can be criticised for lacking a clear protagonist, it is because the audience, accustomed to identifying with other people, are looking in the wrong place. New York itself is the main character in *Style Wars* and only by understanding its component parts, the elements that exist below the mainstream culture, can one truly grasp the city's character.[15]

A more humanistic approach to this subject matter can be found in *Exit through the Gift Shop* (2010) by famed street artist Banksy. Originally rooted in the work of amateur videographer Thierry Guetta, Banksy's film explores the street-art phenomenon through an unexpected case study, turning the story of a movement into the narrative of Guetta's unlikely transformation from documentarian into a prominent (if controversial) figure in the street-art movement. Originally intended as a documentary about street art's twenty-first-century resurgence, based around the material captured by Guetta in the early 2000s, the film had to be completely re-tooled when its original director proved woefully unable to produce competent, or even watchable, content. According to Banksy, the film Guetta produced was so bad that he had to completely reassess his position: 'I realised that maybe [Guetta] wasn't really a filmmaker. That he was maybe just someone with mental problems who happened to have a camera.'[16] To rescue the material, Banksy asked for Guetta's raw footage in the hope that he could re-edit it into something of value. It was at this point that Guetta turned his hand to producing street art of his own, providing the film, which Banksy was now directing, with its new narrative focus.

Rather that re-tooling Guetta's original footage into a *Style-Wars-*esque documentary, as seems to have been the plan, Banksy instead chose to tell the story of Guetta himself, charting how an amateur videographer was able to ingratiate himself into the street art scene and, even more importantly, what he did after he surrendered control of his film to Banksy. Despite lacking any significant artistic talent, Guetta,

---

15 *Style Wars*. Directed by Tony Silver. New York: Public Arts Films, 1983.
16 *Exit Through the Gift Shop*. Directed by Banksy. London: Revolver Entertainment, 2010.

with the help of a large team of paid artists, staged a massive show in Los Angeles in 2008, turning himself, practically overnight, into one of the world's most commercially successful street artists. According to many of Guetta's former subjects, many of whom appear visibly annoyed or offended by Guetta's self-styled rise, their former documentarian was, essentially, over-praised (at best) or a hack (at worst). The art he produced was deeply derivative; and it was principally produced by Guetta's team, rather than the 'artist' himself. In the film, much attention is paid to Guetta's vanity, which is on full show throughout.[17]

And yet *Exit through the Gift Shop* looks fondly at its subject, in spite of the criticisms levelled at him. Banksy drew heavily upon Guetta's original footage and, particularly in the first part of the film, uses it to provide a fascinating insight into street art's renaissance. Nonetheless, the real focus of the film is not the movement itself, but Guetta's attempt to acquire through it the type of external validation he seems to crave and require. Despite his potentially damaging and artistically disingenuous career, it is Guetta's very relatable need for inclusion that sits at the heart of Banksy's film.

By setting aside the need to create an accurate document of the movement's rise, and instead exploring the story of the film's would-be creator, *Exit through the Gift Shop* is able both to surprise and enlighten its audience. The lens through which the movement is viewed is much more personal than might be expected. An unusual (and arresting) life story was used to explore the commercialisation (and possibly the meaninglessness) of an artistic movement, a discussion of arguably greater value than the seriousness with which the subject might have otherwise been treated. In part, *Exit through the Gift Shop* is effective precisely because it suggests that street art might not be as worthy of celebration as its main practitioners believe it to be. Whatever else can be said about Thierry Guetta, he helps to show that the value of art, or an artistic movement, is entirely subjective. Despite failing to produce a documentary about the twenty-first-century version of the street art movement, the makers of *Exit through the Gift Shop* achieve something even more profound.

From a filmmaking perspective, *Exit through the Gift Shop* is an excellent example of how flexibility in the face of reality can lead to the creation of

---

17   Ibid.

documentaries that far exceed their original potential. By accepting that a documentarian should react to circumstances, rather than trying to control or misrepresent them, as many of the films previously discussed in this chapter have done, Banksy's work was able to achieve a greater level of depth and insight than otherwise might have been possible. Events that might have felt like an annoyance or a distraction at the time were instead correctly appreciated for their intellectual and narrative potential. This transformation of perspective even helped to redeem much of Guetta's original footage, turning unusable moments of ham-fisted videography into invaluable character insights. In other words, the nature of the "truth" contained in that film matured significantly.

For the filmmaker-scholar, *Exit through the Gift Shop* should serve as a reminder that they cannot know precisely what type of film they are making until the filmmaking process has concluded; that even the most irrelevant or asinine footage might, if assembled correctly, allow the filmmaker to engage in a more meaningful intellectual discussion than the one they had originally envisioned. Collating the necessary variety of raw material, combined with flexibility in how it is assembled, opens a vast multitude of opportunities.

In many of the examples outlined in this chapter, footage of varying sorts is used in unexpected and novel ways, and these films interact with one another, building upon prior ideas in the genre whilst reacting against others. The authority of Ken Burns's *The Civil War* echoes through *Style Wars*, but with a vastly different set of subjects benefitting from the perceived power of a strong authorial voice. *Nanook of the North*'s semi-staged authenticity is unconsciously mocked by the very different type of authenticity that Banksy injects into *Exit Through the Gift Shop*: one film's B-Roll becomes another film's A-Roll. A deeper truth about the human condition was sought by both *Confessions of a Superhero* and *King of Kong*, but both films ultimately service the need to elicit sentiment and to create entertainment — goals they thoroughly achieve. In each of the examples discussed in this chapter, candidates for the truth have been presented, but each, in its own way, serves as a reminder that those candidates have been constructed with strong authorial voices or editorial agendas.

From an intellectual perspective, you should be prepared to revisit your footage as your project evolves. Indeed, you should be prepared for the creative process to invert your own expectations about the focus of your work. Capture A-Roll and B-Roll, but be prepared to reassess the worth (and classification) of each. By engaging with a wide range of filmic precedent, and by placing your work within the context of its medium, as well as the relevant scholarly literature, your work will be in a position to react not only against the surrounding academic discourse, but a wider environment in which the public is petitioned to invest in innumerable, often manipulative, explorations of the "truth".

7. Choosing Your Filmmaking Equipment

Fig. 11. Watch the next lesson in our documentary-making course. https://hdl.handle.net/20.500.12434/c9b0ef48

Historically speaking, professional-grade filmmaking equipment has long been out of reach for most. High price points, the need for expensive film stock and processing, and the required specialist knowledge proved to be a near-insurmountable barrier for many would-be filmmakers. Radical changes to consumer technology, however, have fundamentally changed this. From the smartphone you likely already own, to more powerful and versatile cameras, there are many options available to you.

In this chapter and video lesson, we discuss the different types of equipment you may wish to utilise for your project. From smartphones

© 2021 Darren R. Reid and Brett Sanders, CC BY-NC 4.0   https://doi.org/10.11647/OBP.0255.07

to dedicated cameras, to tripods and microphones, we aim to provide you with a useful insight into how you can use the tools you already own, as well as those you may wish to acquire, to help you achieve your creative and intellectual goals. From smartphones to DSLRs and more specialised cameras, the potential range of options, at practically every budget level, for filmmaker-scholars is staggering. Whilst technology moves too quickly for this volume to offer an up-to-date guide, comparing and contrasting two case studies should provide you with enough relevant knowledge and context to inform any purchasing decision.

Smartphone Kit ($100–1,000)

Smartphones open up filmmaking to practically everyone. Modern phones (the type which, in all likelihood, you already own) record videos at resolutions of 1080p to 4K. Through the addition of a lavaliere microphone, you can record broadcast-quality sound along with your video. In addition, basic video-editing apps, such as iMovie, even allow you to edit and release a film from within the confines of a single device. For on-the-ground reporting, video journalism, or the creation of more involved pieces, smartphones can open many creative and intellectual doors.

The Kit:

- Camera: your existing smartphone, recording video at a minimum resolution of 1080p.
- Stabilisation: a tripod with smartphone adapter — this can be used to create stable, still footage, or it can be picked up to allow you to go handheld. More advanced solutions, such as motorised gimbals, are also available for smartphones.
- Audio: a lavaliere microphone paired with an older smartphone (acting as your sound recorder — dedicated sound recorders can also be purchased).
- Lenses: lens kits for smartphones are generally inexpensive and may add some additional functionality to your device. These can include macro lens adaptors (to allow your device

7. *Choosing Your Filmmaking Equipment* 77

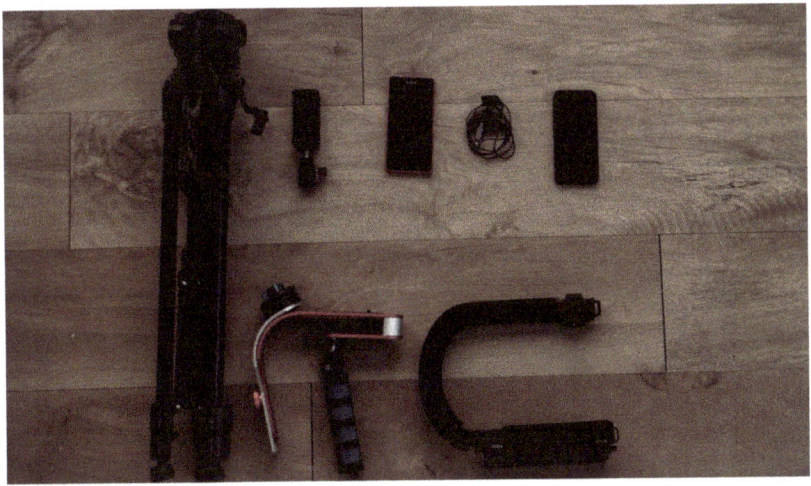

Fig. 12. With only a small additional investment, you can transform the equipment you already own into a basic documentary-making kit. You can utilise your existing smartphone if it is able to capture HD or 4K footage. An older model can be paired with a lavaliere microphone and used as a sound recorder. An inexpensive smartphone adaptor would allow the phone to be connected to a tripod or to one of the stabilisation devices pictured (a gimbal and C-grip). Excluding the cost of the phone(s), the equipment in this setup could be purchased for a total of approximately $120. Pictured, from left to right, top to bottom: tripod, phone holder with tripod adaptor, mobile phone, lavaliere microphone, second mobile phone, gimbal, c-grip.

to focus on objects very close to its lens) or zoom lens adaptors (allowing your device to film subjects that are further away).

- Filmmaking apps: FiLMIC Pro is currently an excellent option for smartphone users. It allows users to control specific settings on their device, allowing it to record footage at 24 framer per second, the same as most traditional film cameras (see chapter eight). In terms of editing, versions of iMovie and Adobe Premiere are both available for a variety of smartphones.

DSLR Kit ($300–5,000)

If smartphones and tablets provide a basic and accessible entry point, affordable consumer DSLRs (cameras with interchangeable lenses) offer filmmakers greater flexibility and the opportunity to capture footage that is of a higher, more cinematic quality. Whilst they are sometimes expensive, older camera models purchased second-hand can offer filmmakers an opportunity to build a comparatively inexpensive kit around a quality piece of filmmaking technology.

Fig. 13. Assembled over time, a DSLR kit's cost can be staggered. This setup was assembled over two years, and cost approximately $800. The camera is a Nikon D5500. It has 18–55mm, 55–200mm, and 50mm lenses alongside a range of filters, a lens hood, and wide-angle and macro adaptors. A gimbal allows for smooth handheld footage, as does a C-grip. A smartphone with a compatible lavaliere microphone helps to round out this kit. Pictured, from left to right, top to bottom: tripod, c-grip, directional microphone, LED light panel, LED filters, focus pull, lens, lavaliere microphone, a pair of lenses, cold shoes, Nikon D5500, lens, mobile phone grip, assorted lens filters, mobile phone.

The Kit:

- Camera: entry-/mid-range DSLRs by Canon, Nikon, Sony (or others) that record video at a resolution of at least 1080p are available for less than $1,500. For budget-minded filmmakers, older camera models, particularly when purchased pre-owned, can help to reduce this cost. At the other end of the spectrum are full-frame DSLRs. These record higher-quality footage than the 'cropped sensors' found in cheaper models, with a price point that corresponds to this increased fidelity. Expect to pay in excess of $2,000 for a full-frame DSLR camera.

- Stabilisation: a tripod and other stabilisers. These will allow you to capture high-quality stationary and moving shots. Stabilisers need not be expensive. A C-grip can provide a versatile handheld option for under $30.
- Audio: your existing smartphone or tablet coupled with a lavaliere microphone. In addition, a directional microphone, which can be connected directly to your camera, will significantly increase the quality of the audio natively captured by your camera.
- Lenses: a range of lenses with variable focal lengths and apertures (also known as f-stops). When building a lens collection, aim to accumulate devices that will offer unique or distinct characteristics. For instance, a lens with a powerful zoom; a lens with a large f-stop; a lens with a wide field of view.

Remember:

- Sensor size: unlike DSLRs, the sensor (the chip onto which focused light is projected) on smartphones is very small. This means that even though a smartphone might record video footage at a resolution of 4K, it will capture much less detail than a DSLR with its larger sensor.
- Dynamic range: smartphones and tablets capture a comparatively limited spectrum of colour compared to most DSLRs. A higher dynamic range means that a camera captures more colours, which can add significant depth to footage, helping to give it a cinematic feel.
- Low-light performance: Cameras with poor low-light performance (a particular problem in smartphones) can add noise, grain, and other undesirable artefacts to footage.

8. Core Methods

Fig. 14. Watch the next lesson in the video series. http://hdl.handle.net/20.500.12434/1956f791

If you are not used to capturing video or making films, as with any new endeavour, starting out can be an intimidating process. But it is also a wonderful and enjoyable adventure, rooted in just a few core methods that can be easily learned and memorised. There is every reason to turn any apprehension you may feel into excitement. Practice, of course, will be required for you to employ these basic rules effectively, but they should allow you, from an early stage, to start capturing competent, usable footage.

Stabilise your Camera

Always use a tripod or, at a push, a monopod — even when going handheld, attach your camera to a support mechanism of some kind.

Avoid touching the camera if it is at all possible, as this can add unwanted movement to your footage. On a small preview screen, or even a mobile phone screen, the amount of shake transferred from your body to your equipment may not be particularly evident. In fact, you can walk away from a shot convinced that you captured a beautiful piece of footage only to discover that, upon review, it is mostly unusable. Luckily, the solution to this is simple: always stabilise your camera.

Use a tripod when capturing stationary shots. When you need to move the camera, use a stabilisation device (such as the C-grip). These will allow you to move your camera without transferring undue amounts of shake to your footage. Remember, your hands move in ways that you are not conscious of, and it is important to counter such movement to ensure you capture high-quality, usable material.

Moving a camera and capturing usable footage is difficult but it can be done using relatively inexpensive equipment and a *lot* of practice and patience. Rather than planning complicated camera moves during the early stages of your filmmaking career, you will be better served if you focus your energies elsewhere. Practice creating really stable, well-composed shots which can communicate your ideas as well as any movement of the camera. Remember, cutting from a well-composed wide shot to a considered and intimate close-up can be just as effective as moving the camera towards your subject. If in doubt, keep your camera stationary. Practice and experience will allow you to begin experimenting with moving your camera in due course.

Focus your Camera on your Subject

Whether you are using an expensive DSLR or the camera on your smartphone, always focus on your subject. In the case of a human being, focus on their eye — there is no point in focusing on someone's nose when the eye is the window to the soul. Unfocused shots can remind an audience that they are watching a film, breaking the immersion of the movement, and destroy the aesthetic quality you sought to create with your composition. A distracted audience is a disengaged audience; your viewers demand (even if they are not conscious of it) well-focused shots.

Remember, every time you move your camera (or when a subject moves within your frame) you will need to refocus it. If you are moving your camera, or if you are photographing a moving subject, refocus your shot for every new take. In the case of smartphone cameras and modern DSLRs this can be as simple as touching a point on the screen. There is nothing worse than composing a perfectly stable shot only to find that your point of interest is out of focus when you review your footage at a later date. Otherwise usable footage will be rendered unusable by such an oversight.

After focusing, particularly when using lightweight equipment, such as a DSLR or smartphone, give your camera a moment to rest so that any residual motion, transferred from you to the equipment, has had an opportunity to dissipate.

Compose your Shots

Stability and focus make a shot bearable — shot composition is what makes it worthwhile. It does not matter if you're shooting on a $10,000 camera or a comparatively inexpensive smartphone, careful and considered composition adds aesthetic value and, if used correctly, intellectual beauty to your work. Even a shot compromised by poor technology can be beautiful and emotive if time has been spent to compose it with care.

Obviously, there are instances when shot composition is not as important as it otherwise might be. The footage of planes flying into the Twin Towers on 9/11 videos are made no weaker by the lack of thought placed into their composition. But unless an event is fundamentally extraordinary, unusual, or dramatic (or there is an obvious reason for an audience to forgive poor composition), poorly-composed shots are not likely to be as effective as they otherwise could be.

To compose effective shots, you should utilise the 'rule of thirds'. The human eye does not find images in which a subject is placed directly at their centre to be consistently satisfying. Instead, the eye appreciates an image that is imbalanced in some way. The 'rule of thirds' is, contrary to its name, a piece of compositional guidance rather than a definitive law which must be followed at all costs.[1] There are many pieces of visual art that do not conform to this grid and which would not have been

---

1   For an early, pre-photographic description of the rule, see John Thomas Smith, *Remarks on Rural Scenery* (London: Nathaniel Smith, 1797), pp. 15–17.

improved by its use.[2] That being said, its use throughout the history of film has created a learned appreciation for it among modern audiences. To use the rule of thirds is to appeal to the subconscious expectations of one's audience. Mentally project the 'rule of thirds' grid over practically any film and see how the filmmaker uses the 1/3 axes, both horizontal and vertical, to compose their images. This consistency of approach means that most audiences associate such compositions with high-quality productions. Such compositions, in other words, feel right.

This is how it works: divide your viewing area into thirds, both horizontally and vertically, as seen in Figure 15 This will create a grid: memorise it and see it everywhere you look. Project it onto the world around you. Now impose that grid over a photograph, as seen in Figure 16 Whilst there is nothing egregiously offensive with the photograph in Figure 16, it is not particularly cinematic. The subject is centred, presented in a non-dynamic and uninteresting way. By using the 'rule of thirds' instead, that same subject can be framed in a more visually dynamic way.

The 'rule of thirds' allows you to present your subjects with implied tension in the composition. The eye prefers images that are not balanced, unless that balance serves a deeper aesthetic, intellectual, or symbolic purpose. Experiment by photographing people or other subjects whilst employing this principle.

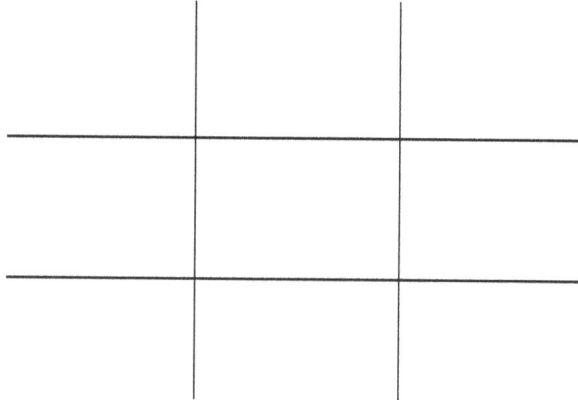

Fig. 15. The 'Rule of Thirds' grid is frequently used to shape filmic compositions.

---

2   Bert Krages, *Photography: The Art of Composition* (New York: Allworth Press, 2005).

Fig. 16. This photograph makes little use of the grid, its subject having been centred without regard for the ways in which the axes of the grid might add tension to the frame.

The image below (Figure 17) utilises the 'rule of thirds' and, as a result, implies a relationship between the subject and their surroundings that was not previously present in the original photograph (see Figure 16). The substantial space to the side of the subject provides them with space into which they can look or move.

Fig. 17. By moving the subject off-centre and lining them up along one of the 1/3 axes, a degree of tension and imbalance is added to this composition. There is now space into which the subject can look and there is a clearer sense of compositional clarity. Even in a still photograph, the viewer is primed to expect the subject to move from left to right, through the vacant space within the frame.

In terms of interviews, a good rule of thumb is to align your subject with the upper intersection of one of your vertical and horizontal axes, as in the following image (see Figure 18). This will help to place your subject in your frame in a way that feels familiar and well-composed to most of your audience.

Once you start experimenting with the rule of thirds, you will find that your compositions begin to develop their own dynamism, feeling more deliberate and effective in their composition. Of course, there are times when this rule can and should be broken, but learning and understanding the rule will help you to do so effectively. Again, practice is the key to getting the most out of this technique. Next time you photograph a person or scene, line up different elements in your shot with the 'rule of third' axes and experiment with the results.

Fig. 18. For interviews, try lining up one of your subject's eyes with one of the intersections of the upper axes, as seen in this image.

Plan to Capture Contextual Footage

In his mammoth fourteen-part film series on the history of film, *The Story of Film* (2014), Mark Cousins places a huge amount of material, which might normally be considered B-Roll, front and centre.[3] Despite employing a wide range of interviews, Cousins populates much of his

---

3   *The Story of Film: An Odyssey.* Directed by Mark Cousins. Edinburgh: Hopscotch Films, 2011.

series with footage of urban environments (typically related to the locales he is discussing), shot from numerous angles and cut in a way that allows the sequence of images to reflect the themes in his commentary. In so doing, Cousins shows how footage of physical spaces can speak to deeper themes being discussed in documentaries. What might have comprised only a fraction of the shooting time in a traditional TV-style documentary instead has attention lavished on it.

Everything you shoot has the potential to define your film. Do not assume that any of your footage will prove to be of lesser value to you. Interviews and other set-piece moments are naturally going to be important, but carefully photographing the environment and other incidental pieces of footage (B-Roll) can open up significant options once you enter the post-production process.[4] Indeed, B-Roll hardly feels like an appropriate label, considering how flexible this footage is, and how centrally it can be used. For the purposes of this discussion, the phrase B-Roll will not be used again. Instead, it will be referred to using a less pejorative label: contextual footage.

As much attention should be paid to capturing contextual footage as is paid to filming interviews or other important set pieces. Interviews may very well be the foundation of your film, but you will likely need at least some shots of your subjects' context to serve as connective tissue. Contextual footage can help to place your interviewees in an environment that reflects or contrasts with their spoken ideology. In other words, the ways in which you place (or do not place) your subject into context helps to inform how your film is read and understood by its audience.

For example, in a documentary about the history of law, you might shoot footage of your interview subject standing in the entrance of a courthouse. If you shoot wide (at a distance from your subject) you can frame them so that they are dwarfed by the size of the court, the physical manifestation of the law's power, about which they are an expert. In order to capture the scale and scope of the court (and what it represents), you would probably have to pull your camera so far back that your subject would become lost in the resultant frame. Thus, in order to make this shot work, it would have to be a part of a sequence: a

---

4   *The Story of Film: An Odyssey*. Directed by Mark Cousins. Edinburgh: Hopscotch Films, 2011.

wide shot could give way to a mid-shot in which the subject is somewhat more identifiable. A third shot could then move closer still to the subject; the camera would now be close enough that the audience can clearly identify the subject. The film could then cut to the subject, sat on a chair indoors. In this example, the subject has been placed in context, dwarfed by the institution in which they serve. The resultant mid and close shots ensure that the audience is aware that they are viewing a subject, indiscernible in the first shot, in the context of their surroundings and life's work.

Alternatively, that same footage could be sequenced in reverse. The interview might have a cold start (no lead-in footage) but, as the interview nears its conclusion the film would then begin to (literally and symbolically) move away from the subject. In this example, the sequence of shots would be: the subject being interviewed; the subject in the door of the courthouse (close up); the subject in the door of the courthouse (mid shot); the courthouse (wide shot). In this sequence, the camera (and thus the audience) moves away from the subject — clarity gives way to obscurity, rather than the move towards greater intimacy with the subject implied by the original assembly. The same footage, assembled differently, can thus provide a substantially different meaning — contextual footage, in both instances, plays a key role in achieving either effect. With sufficient contextual footage, numerous opportunities, many previously unimagined, emerge in the post-production phase. Without that material, the potential to experiment with the assembly of the film is substantially reduced, if not eliminated entirely. By documenting the subject's context (and the subject in context), you will greatly broaden the size of your visual alphabet.

You might also construct visual montages from contextual footage that echo or rhyme with interview dialogue (or commentary tracks). In a film about homelessness, for instance, an anecdote about life on the streets during the winter months might be illustrated with a visual sequence showing shots of a city, shot low (the vantage point of someone sitting or lying down on the pavement) to create a type of accompanying visual essay: pools of stagnant, freezing water; feet and legs passing in front of the camera; small groups of affluent young people chatting convivially, happily soaking in their surroundings, shot from a distance. Depending on how you photograph this contextual footage, and the order in which you sequence it, it will take on any number of different meanings which will add intellectual and aesthetic depth to your work.[5]

There is significant creative and intellectual value in treating your contextual footage with as much weight as you treat your A-Roll. By paying attention to one's environment and endeavouring to film it in a way that captures its vibrancy and contradictions, new themes can be brought out and discourses deepened. Achieving this, however, will require you to make a concerted effort to document spaces as much as you document people or events. Consider both practical and symbolic uses for the environmental footage. Practically speaking, such footage can lead into and out of interviews, or provide a visual cue over which a commentary track can run. But symbolically, a space can serve a much deeper purpose when it is photographed and explored on screen. The example of the law historian in front of the courthouse merely touches upon that potential.

Contextual footage can tell a story about a space, creating new truths that speak to the themes and subtexts linked to, or at odds with, those explored explicitly in your film. A sequence of shots, moving towards, away from, or about a space can help to create an effective narrative or thematic frame. Each shot of the environment, each cut, should serve to develop that frame, bringing out the specific details of the narrative or topic. This might be accomplished by gradually positioning the camera closer to a building, as in our earlier example, bringing the audience closer to a subject or some symbolic detail in the environment. Or the camera might move in a less organic way, cutting from one detail to another without particular attention being paid to how the shots relate to each other spatially. In the courthouse example, this might mean cutting between different details carved into the building's facade. On a medieval church, such shots might focus on the religious iconography carved into the structure, gargoyles and stained-glass motifs.[6]

Capturing copious and considered contextual footage alongside your A-Roll will provide you with many options when it comes to

---

5   Dancyger, *The Technique of Film and Video Editing*, pp. 16–22.
6   *Los Angeles Plays Itself*, by Thom Andersen, is a masterclass in its own right on the use of contextual footage in order to tell the story of a space. See *Los Angeles Plays Itself*. Directed by Thom Andersen. New York: The Cinema Guild, 2004.

your film's assembly, but there is a vast array of precedent that will help to inspire the possibilities open to you; examples that you can follow, discard, build upon, and react against. Consider, for instance, the opening of Woody Allen's 1979 dramatic film, *Manhattan*, in which shots of Manhattan Island are cut together to the lackadaisical opening of George Gershwin's 'Rhapsody in Blue'. The sequence is comprised entirely of contextual footage and functions as a type of short film in its own right, a little poem about life in the bustling heart of New York. The sequence is remarkably effective considering the relative simplicity of its component parts — contextual footage, a piece of music, and an audio commentary. Whilst documentarians may not see value in emulating Allen, there is much they can learn by studying and reacting to this sequence.[7]

Shoot Longer Takes

As an example, assume that you need a thirty-second shot of a building's exterior. You set up your camera on a tripod, focus it on the front the building, and press record. How long shout you leave your equipment recording?

Obviously, thirty seconds is the bare minimum duration for such a shot, but, ideally, you should leave your equipment to record for quite a bit longer. Contact with your hand (when you hit record) may start the camera shaking slightly and it may take several seconds for the camera to become entirely stabile again. Perhaps more importantly, you may come to realise during the editing process that what you actually needed was forty-five or sixty seconds of footage. The solution is to take longer shots as a matter of course, ensuring you have maximum flexibility during the editing process.

Do not capture footage that only meets your minimum spec. As a rule of thumb: double what you need and then add a fifteen-second 'leader' at the start (to compensate for any camera shake). Thus, if you need thirty seconds, shoot for 1:15 (15 seconds, plus 30 x 2). This should ensure that there is at least thirty seconds of usable footage. If time is not a factor, quadruple your minimum requirement and add fifteen seconds.

---

7 *Manhattan*. Directed by Woody Allen. Los Angeles: United Artists, 1979.

In many, probably most, cases, however, you will not know how much footage you actually require. There is no simple rule of thumb should you begin filming a shot that you had not previously anticipated, but you should endeavour to capture enough footage to ensure that your footage can be used in a wide range of ways during the editing process. Ten to fifteen seconds might feel like a more than adequate amount of footage when you are in the field, but during the editing process it will severely limit your options. One minute and thirty seconds might feel like an excessive amount of time to record, for example, a building's exterior, but such a long shot will give you many possibilities that a shorter shot would not.

If you find an interesting scene, set up your equipment and begin recording. If something is unfolding, capture the entirety of that event — and then keep recording. You might not realise it at the time, but the camera may capture an interesting after-effect. If you are shooting a car, perhaps it will lift some leaves into the air; off-site, you may realise that it is the shot of the leaves blowing in the car's wake that is the most visually or symbolically dynamic part of the footage you captured. This may not have been evident to you when you captured the footage in the field.

In other words: shoot more than you require. Never shoot the bare minimum.

Take Control of your Camera's Settings

Whether you are using a smartphone or a DSLR or a pro-camcorder, your equipment will provide you with at least some control over its operations. It can be tempting, particularly for first-time filmmakers, to simply set their camera on automatic. Doing so, however, means that you will forgo a significant amount of control. It might also result in footage that, for one reason or another, does not conform to how most people expect modern cinematic footage to look or feel (see chapter nine).

As a rule of thumb, you should film at twenty-four frames per second (fps). This is the frame rate at which most films are shot and, as a result, *feels* correct. A century of cinema has conditioned us to expect a certain look from movie footage. Consider the negative reaction surrounding the 48fps (high frame rate) release of Peter Jackson's *The Hobbit* (2012–2014) trilogy. Despite costing hundreds of millions of dollars, critics complained that the frame rate, whilst smooth, made the film look and feel cheap.[8] What they meant was that the increased smoothness of the frame rate made the film feel un-cinematic. It was too realistic and, as a result, audiences were disturbed and taken out of the moment — they found it more difficult to suspend their disbelief. A frame rate of 24fps will help to provide a subtle cinematic feel to your film. It will almost certainly not be noticed or appreciated by your audience, but its absence might.

If you are using a DSLR there are a number of other settings that will allow you to capture footage that feels even more cinematic — see chapter nine for a complete breakdown of how to set up your camera.

If you are using a smartphone or tablet, there are a number of apps that will allow you to gain greater control over your camera's settings. Currently, FILMiC PRO offers iOS and Android users the ability to change the camera's frame rate and method of recording sound, whilst introducing separate controls for focus and exposure. These features will empower you to capture higher-quality footage. The use of manual control is, of course, more time- and labour-intensive, but the results easily negate this.

---

8   For examples, see Jen Yamato, 'The Science of High Frame Rates, Or: Why "The Hobbit" Looks Bad at 48FPS', *Movieline*, 14 December 2012, http://movieline.com/2012/12/14/hobbit-high-frame-rate-science-48-frames-per-second; Vincent Laforet, 'The Hobbit: An Unexpected Masterclass in Why 48FPS Fails', *Gizmondo*, 19 December 2012, https://gizmodo.com/5969817/the-hobbit-an-unexpected-masterclass-in-why-48-fps-fails; Anthony Wong Kosner, 'The Reason Why Many Found The Hobbit an Unexpectedly Painful Journey', *Forbes*, 11 January 2013, https://www.forbes.com/sites/anthonykosner/2013/01/11/the-reason-why-many-found-the-hobbit-at-48-fps-an-unexpectedly-painful-journey/#6f2143ba31cf. For alternative perspectives, see Hugh Hart, 'The Hobbit is Insanely Gorgeous at 48 Frames Per Second', *Wired*, 12 December 2012, https://www.wired.com/2012/12/hobbit-movie-review-48-fps/ and Jacob Kastrenakes, 'The Hobbit's Vision for the Future of Cinema Looks Awful, but it Might Just Work', *The Verge*, 19 December 2014, https://www.theverge.com/2014/12/19/7422633/hfr-might-work-even-though-it-looks-really-awful

9. Settings, Lenses, Focus, and Exposure

Fig. 19. Watch the next lesson in the video series. http://hdl.handle.net/20.500.12434/92a4bc2b

There are a number of settings and features on your camera with which you should familiarise yourself. As much as possible, you should move away from the automatic mode on your camera and begin setting it up to accommodate the conditions in which you find yourself. As much as possible, this chapter will continue to provide practical, actionable information. There is much more to be said about lenses and how they function, but that information is not required in order to utilise your lenses effectively. Remember, there is much that can be learned beyond this text about these topics, but the information here should prove sufficient to facilitate a quick and effective transition into the field.

© 2021 Darren R. Reid and Brett Sanders, CC BY-NC 4.0   https://doi.org/10.11647/OBP.0255.09

This chapter will provide you with the knowledge needed to quickly begin utilising your camera to its best potential. The first section contains the standard camera settings that you should use in order to capture footage that feels analogous to film (cinematic). The second, third, and fourth sections build upon this by providing information and techniques that will allow you to begin to stylise the footage you capture.

Camera Settings

If your camera has the option, you should adjust the following settings as closely as possible to the following specifications.

- You should set your frame rate to 24fps.

This is the standard frame rate that is most closely associated with the look and feel of celluloid. A lower frame rate can give your video a choppy feel which will likely make your audience feel uncomfortable. More than this and an image can become too smooth and will start to feel like the cheap video on which television shows were frequently shot in the 1980s and 1990s. There are, of course, exceptions. If you have a camera capable of shooting at, say, 60fps, then you will be able to slow down your footage to a fraction of its normal playback speed, capturing super-smooth slow-motion footage. As a rule, you should only shoot at 60 or 120fps (etc) when you want to capture such slow-motion sequences.

- Your shutter speed should be 1/frame rate x 2.

This is only applicable if you have a camera that allows you to control your shutter speed (such as a DSLR). If you do, apply the above formula as closely as possible. If you are shooting at 24fps, your shutter speed should be 1/48 (1/24x2) — or as close to that as your camera allows (1/50 is a common setting on most DSLRs). If you are shooting at a high frame rate for slow-motion shots, such as 60fps, the shutter speed becomes 1/60x2 — or 1/128.

- White balance can be used to change the hue of your footage. Essentially, this controls the 'temperature' of your image. A low temperature gives your image a blue tint, and a warm temperature gives it a yellow tint.

For everyday shooting, setting your white balance to automatic should be sufficient. But if you wish to give your footage a specific look, experiment with the different settings on your camera. Your camera, if it allows for white balance control, is likely to contain settings for different locales — for instance, there is likely to be a white balance pre-set for shooting under florescent light, a pre-set for shooting in cloudy conditions, and a pre-set for shooting in bright sunlight. To stylise your footage, try using a white balance pre-set that it is not intended for the conditions in which you find yourself.[1]

Alternatively, shoot using the appropriate pre-set and then colour-grade your footage in post-production to give it the desired effect. Approaching the stylisation process in this way means that a neutral version of your original footage, should you ever need it, will be available to you. If you are shooting using more than one camera, pay particular attention to the white balance on both cameras to ensure that they are capturing footage that is comparable. Particularly when using cameras by different manufacturers, it may be necessary to set the white balance on both cameras manually to ensure a consistent temperature profile between your shots.

Lenses

It is important that you understand some basic rules about how lenses work. Even if you are using a fixed-lens camera or a smartphone, you should have some grasp of how lenses capture footage in the way that they do, to ensure that you can anticipate how your equipment will perform in different situations and conditions.

The focal length of your lens is measured in mm — the smaller the number, the wider the shot. An 18mm lens would capture a wide view (zoomed out) of a scene, whereas a 200mm lens, looking at the same area, would instead capture a close-up (zoomed in).

Aside from zooming in, however, the focal length of your lenses also affects the type of image that your camera captures, particularly with regard to how close background and foreground objects appear in relation to one another. A lens with a small focal length will preserve

---

1   See Hugh Fenton, *Cinematograph: Learn from a Master*, YouTube, 27 April 2012, https://www.youtube.com/watch?v=KwtpJ3T8eK4&t=7s

Fig. 20. Two subjects standing approximately eight feet apart, photographed using an 18mm lens. Note how small many of the background details are. All rights reserved.

the sense of distance between the foreground and background, whereas a lens with a longer focal length will compress (squash) the distance between them. Consider Figures 20, 21, and 22. The subjects remain stationary; only the lenses have been changed. Note how the spaces between the two subjects is compressed, as the focal length increases.

Fig. 21. The same two subjects, standing in the same positions, photographed using a 50mm lens. Note how the background subject now appears much closer to the foreground subject. Note also how the background details have increased in size. All rights reserved.

9. Settings, Lenses, Focus, and Exposure   97

In Figure 20 the two subjects have been photographed using a lens with a focal length of just 18mm. In this image, the foreground subject is significantly larger than the background subject. In Figure 21 the same two subjects were photographed standing in the same locations, but using a lens with a 50mm focal length. In this image, the background subject seems to be much closer to their counterpart in the foreground when, in reality, they have remained stationary. In Figure 22, which was shot on a lens with a focal length of 200mm, the background and foreground subjects appear to be almost the same size. By changing the type of lens being used to capture this scene, the resultant compositions produce radically different effects.

Fig. 22. When photographed in 200mm, the background subject (upon whom the focus has now been pulled) appears very close to the foreground subject. Also note how close the environmental background details appear relative to our subjects. The space in this frame has been severely compressed. All rights reserved.

A lens with a large focal length compresses the distance between objects in the foreground and background of your footage. This means that, aside from zooming into a scene, a 200mm lens will bring distant background objects much closer to the foreground. Compare the backgrounds of Figures 20 and 22. Note that in Figure 20, there are buildings in the distance but they appear very small in this composition. In Figure 22 on the other hand, the same buildings now appear much larger. This creates the impression of compressed space.

Lenses with a small focal length can also distort facial features, adding subtle (and sometimes not-so-subtle) distortion and bulge. In contrast, a lens with a larger focal length, say 200mm, will tend to flatten facial features. A lens of about 50mm captures images that produce a reasonable approximation of what is seen by the human eye.

Stylised Focus

You should experiment with focal lengths to create more beautiful or symbolically rich imagery. Shallow focus, where only a part of the shot is in focus, often produces aesthetically beautiful shots which serve to direct the viewer's attention to a specific location within a frame. If a person is filmed in shallow focus, the background around them will typically be so indistinct that the viewer will have no choice but to direct their attention fully towards the subject. In contrast, a deep-focus shot, one in which every part of the frame is clear and discernible, can more effectively place a subject in context.

To achieve shallow focus, you will generally need a lens with a large aperture. The aperture is the hole through which light enters the camera. The wider the aperture, the shallower the focus. The aperture size is measured in f-stops. The lower the f-stop, the larger the aperture and vice versa. An f-stop of 1.4 would allow a lot of light into your camera, but would give you very shallow focus. An f-stop of 3.5 will give you a shot in which most, but not all, of the frame is in focus. An f-stop of 8 will let in a comparatively small amount of light, creating a frame in which much of the detail will be sharp and clear. Most consumer cameras, and the kit lenses that come with most DSLRs, have a reasonably large f-stop, enough that some areas of a shot will be out of focus, but not large enough that you will be able to achieve a highly stylised, shallow-focus look. To achieve this, you should supplement your camera with a lens which possesses an f-stop of 2.8 or 1.8.

For filmmaker-scholars, functionality must trump style; it is important that a transient moment is captured in a usable form. Visual beauty is desirable, but not essential. A shallow focus might help to stylise your footage, but the effort and time required to adjust your focus might result in your failing to capture a significant, but transient moment — and it is better to capture imperfect footage of a rare event than beautiful footage of something inconsequential. Stylised footage can look beautiful — but it can also distract and is generally more time-consuming to achieve. It should be employed with care and consideration.

Many types of digital cameras will not allow for the capture of a particularly shallow focus, but there are ways to force a limited version of the effect. One way to force a shallow effect (particularly on smartphones) is to shoot a scene with a foreground object that is much closer to your camera than the main subject of your frame. For example: place your camera on the ground, a few inches away from a blade of grass. Focus it upon your subject (which should be some distance from the camera), and the blade of grass will blur. The result will be an image in which your subject is in focus, but an out-of-focus foreground object adds some stylisation to the shot. Conversely, the same setup would allow you to focus on the foreground object (in this example, a blade of grass), forcing the background to blur. This approach is quite limited, however, and requires you to think carefully about setting up this type of shot for any type of practical application. If shallow-focus stylisation is something you wish to achieve with regularity, a DSLR with an appropriate lens will be a much better long-term solution for your needs.

You should be wary, however, about sacrificing your composition for the sake of some lens blur. Whilst shallow focus, when used correctly, can certainly add value to a production, it can also be distracting if it is used gratuitously. If, when composing a shot, you recognise an opportunity to use shallow focus effectively, then experiment to see what the overall effect will be. But remember — overall shot composition is far more important than adding some lens blur.

Exposure

Exposure is related to focus, thanks to the light-gathering function of the camera's aperture (f-stop). If a piece of footage is overexposed, parts of your frame will lose detail. 'Burning out' occurs when a camera no longer records information in over-lit areas; where there should be detail and a gradation of colours and shade, the camera will instead only record an area of white without detail. Conversely, underexposed footage stops recording detail in the shadows. In under-exposed footage, parts of a frame become black holes with no discernible nuance or structure, in much the same way that over-exposed sections become white splodges with no detail.

There are two main ways to control your exposure — the size of your aperture (f-stop) and your ISO setting. As already mentioned, the smaller the f-stop (and thus, the larger the aperture), the more light is admitted. This allows you to capture footage with a shallow focus, but on a sunny day you might well find that your image is easily over-exposed as a result. To compensate for this, adjust the size of your f-stop. This will reduce the amount of light entering your camera as well as deepening the focus of your shot (this may be an unwanted side-effect if you are hoping to achieve a shallow focus). Alternatively, you can reduce your ISO, adjusting it until the image is no longer over-exposed. In doing so, however, you might reach your camera's lowest ISO limit (typically 100) but still find that your footage is over-exposed. At this point, you will either need to close up your aperture (and accept that you will not be able to capture a shallow-focus image) or apply a neutral density (ND) filter. These simple devices cut down the amount of light entering the camera, allowing for wider-aperture (lower f-stop) settings to be used in bright or sunny situations. They are typically inexpensive and are widely available. If you intend to capture shallow-focus footage using a DSLR in daylight conditions, an ND filter will be an essential purchase.

For most consumer cameras and smartphones — those without any control of the size of an aperture (f-stop) — exposure will be controlled exclusively via your camera's ISO settings. If you are using such a camera, you should keep at least one eye on your ISO and be prepared to adjust it if parts of your image are either too bright or too dark.

As a rule, you should keep your ISO as low as possible. Increasing your ISO can introduce 'noise' (grain and other visual artefacts) to your footage, reducing its quality. How high you can push your ISO before noticeable amounts of 'noise' appears will depend entirely upon your camera. On some cameras, pushing your ISO beyond 800 will result in a marked decrease in the sharpness of your image and the amount of noise that is visible. In other cases, particularly on newer cameras, the ISO can be pushed significantly higher before the footage quality begins to noticeably degrade.

9. Settings, Lenses, Focus, and Exposure

Experimentation is the key to understanding the usable threshold of your camera's low-light capabilities. As a rule, we try to avoid pushing our equipment beyond an ISO setting of 1600. After this point, the image tends to get noticeably noisy to the point of distraction and footage *can* become unusable (although some newer DSLRs have significantly improved their low-light capability). It is also worth keeping an eye on your ISO level when you are in well-lit conditions. Try to keep your ISO as low as possible to avoid adding unnecessary noise to your footage.

Problematically, most consumer cameras struggle in low-light conditions. This means that as you increase your ISO level, noise is unavoidably introduced to your footage. The higher your ISO, the more noise enters your shots. This can make footage captured in low-light conditions significantly inferior to the footage you capture in well-lit conditions.

Though frustrating, this a reality to which you can adapt. Smartphones, for instance, tend to have comparatively poor low-light capabilities. Properly stabilised and focused, smartphones can capture quality footage but, in low-light conditions, footage that would otherwise have been clear and impressive can take on a low-resolution look and feel. Sound planning (shoot during the daytime in well-lit conditions) can make a big difference. Plan your shoot so that you avoid, as much as possible, forcing your equipment to work under conditions that will produce poor-quality results. Experiment with your equipment so that you become familiar with its limits, quirks, and capabilities. When shooting, we have repeatedly come up against the low-light issue. In fact, new cameras that can handle low-light conditions are of special interest to filmmakers for this very reason. When making *Looking for Charlie* we utilised a Nikon D3100 — out of date even when we acquired it, it was, nevertheless, able to capture usable footage. However, its low-light capabilities were very poor and, as light levels faded, so too did the quality of the footage it captured.

It is possible to shoot in low-light conditions, even with humble equipment, but, typically, the results of shooting at night with a cheap or non-specialised consumer camera will be of a poor quality — even when the camera operator understands the quirks and limitations of the technology at their disposal. But, if you happen to find yourself in a position where you have no choice but to shoot in low-light conditions with crude equipment, shoot anyway.

Get your camera out, stabilise it, frame your shot, focus, and start recording. The very worst that can happen is that you get unusable footage — but you *might* get something that is usable. Do not build your shoot around such endeavours, but if an opportunity presents itself, and there is no other time- or resource penalty for making the attempt, doing so is worthwhile. Even a noisy shot can work in the correct context.

Summary

- Shoot at 24fps with a shutter speed of 1/50 or 1/48.
- The longer your lens's focal length, the greater the zoom.
- The longer your lens's focal length, the shallower the space.
- The smaller your lens's f-stop, the shallower the focus.
- The smaller your lens's f-stop, the more light will enter your camera.
    - A f-stop of 2.8 or lower should be sought for stylised, shallow-focus footage, with 1.8 or lower being the better solution).
- Keep your ISO as low as possible.
- Increase your ISO to increase the sensitivity of your camera's sensor to light.
- Increasing your ISO can introduce 'noise' to your footage and degrade its overall quality.

10. Composing a Shot — Tips and Techniques

Fig. 23. Watch the video lesson on shot composition. http://hdl.handle.net/20.500.12434/18da6176

Just as there are grammatical rules that govern how we write, so too are there grammatical rules that govern how we film (and process) visual information. Composition is a powerful tool, allowing filmmaker-scholars to communicate core ideas and themes without having to articulate them directly. These techniques can also be used to create shots and sequences that appeal to your audience's learned appreciation for the grammatical conventions more than a century of cinema have instilled within them.

© 2021 Darren R. Reid and Brett Sanders, CC BY-NC 4.0   https://doi.org/10.11647/OBP.0255.10

The 'rule of thirds' is an important compositional rule, but there is more to the creation of an effective frame than this rule alone. Head space, looking room, and camera placement will all have a significant impact on the shots you are framing and the impression they make upon your audience. Of course, there are always times when the rules in this chapter should be broken — but even if you choose not to adhere to these rules, understanding them will assist you in breaking them in the most effective ways possible.

Head Room

How many times have you handed your camera to someone to capture a special moment or meeting, only for them to return it with the top of someone's head missing? This is bad composition for obvious reasons, but there is more to 'head space' than simply ensuring that no one is photographically decapitated.

Fig. 24. The subject's head is pressed against the top of the frame, giving the shot an unsatisfying feel.

In Figure 24, much of the subject is visible but their head is touching the top of the frame. Even though the top of their head has not been cut off, the framing of this image feels awkward, as if the subject is being confined by the frame. There are, to be sure, instances when a filmmaker might do this deliberately, but for a standard interview, such a shot might convey an inappropriate impression to viewers. The shot is not zoomed-in enough to be stylised, nor is it far enough away to place the subject comfortably within the frame.

A better way to compose this same shot would be to adjust the camera's height, providing a degree of space between the top of the subject's head and the top of the frame. This type of framing places the subject carefully within a field of view without giving the impression that they are trapped within an enclosed space. Head space should not be excessive, however, as seen in Figure 25.

Too much head room can leave an audience feeling similarly dissatisfied with the shot. If more than a third of the frame is given to headroom, a subject can feel lost amidst their surroundings; it is spatially and visually unclear. Headroom should, then, not draw attention to itself — either as a result of its absence or because of its overabundance (see Figure 26).

Fig. 25. An over-abundance of head room is similarly unsatisfying to the eye. All Rights Reserved.

Fig. 26. A small space between the top of the head and the top of the frame, however, feels appropriate.

Fig. 27. A lack of looking room makes a frame spatially unclear.

10. Composing a Shot — Tips and Techniques   107

Fig. 28. Despite the subject not having moved position, the addition of looking room makes greater visual sense.

Looking Room

Like head room, looking room is one of those compositional rules that audiences unconsciously demand. Looking room is all about achieving an intuitive, spatially clear shot — in this case, providing space into which a subject can stare, or look. Consider the shot above in Figure 27The subject is looking to the left of the shot but their face is pressed up to the edge of the frame. Despite the fact that we know there is space into which the figure must be able to stare, the composition of this shot does not communicate that clearly to the audience. To imply distance between the subject and their surroundings, the filmmaker must include distance in the frame: a space between the subject and the edge of the frame. Figure 27 should thus be reframed to provide distance into which the subject can stare, as per Figure 28. This creates a scene that is compositionally and spatially clear.

The 30° and 180° Rules

Speaking of spatial clarity, capturing an object or subject from multiple angles offers many possibilities when it comes to editing. One could, for instance, set up multiple cameras in an interview situation, allowing the filmmaker to cut between different shots of the same subject. Shooting the same scene from multiple angles is called 'coverage' and the more coverage you capture, the more freedom you will have during the editing process. Coverage of the same event (such as an interview) will also allow you to cover mistakes or other errors captured by any one camera by cutting to a different angle.

Capturing a significant amount of coverage requires you to learn some important compositional rules. The first is the '30° rule', which stipulates that at least 30° of separation must exist between camera angles that you intend to cut together. If a film cuts between two cameras that are not at least 30° apart, the audience will likely realise that a cut has been made. As a result, they will remember that they are watching a film and the immersion of the moment will be broken.

At least 30° of separation should sit between shots that are to be edited together. See Figures 30 and 31.

The '180° rule' will similarly help you to shoot footage that will be spatially clear. In a conversation between two people, such as an interviewer or interviewee, imagine an axis drawn between them, as in Figure 29.

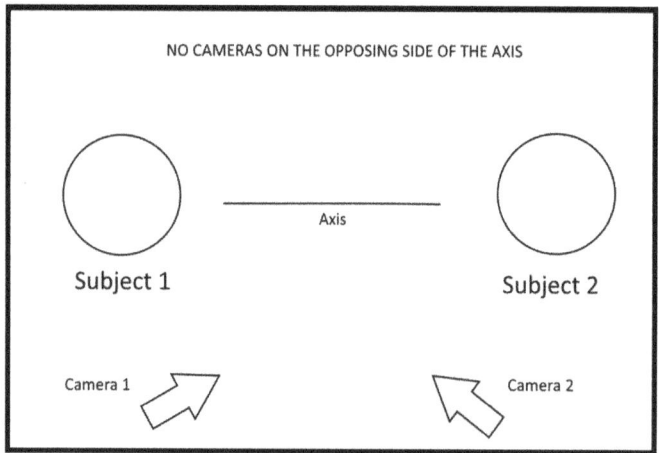

Fig. 29. When shooting an interview, cameras should be positioned on one side of the 'axis' only.

All cameras recording this conversation should be placed on the *same* side of this axis. If cameras are placed on opposite sides of the axis they will create a spatially confusing scene in which both subjects appear to be facing the same direction, not one another. Whenever you are in a situation in which two objects or subjects are meant to be shown facing one another across different cuts, the 180° rule should be rigorously observed.

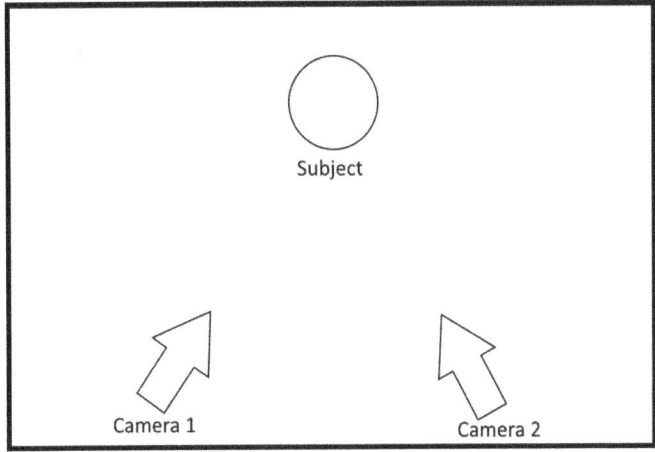

Fig. 30. Two cameras photographing the same object.

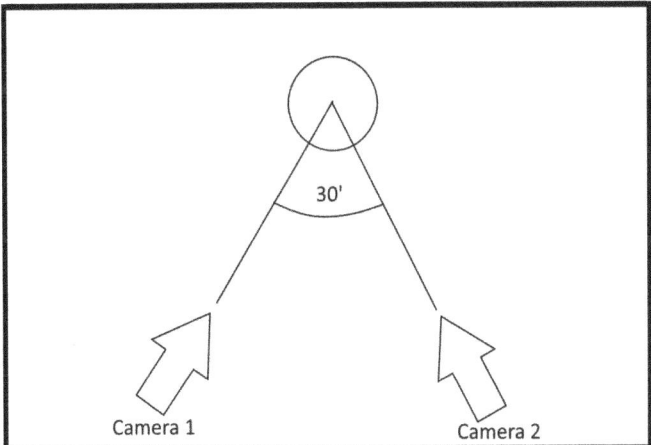

Fig. 31. The cameras should be at least 30° apart, or the audience may become aware of the cut between these different angles.

Fundamentals

Over the course of the past two chapters, you have learned the fundamentals that will allow you to begin shooting effective footage. There is, of course, a lot more than can be said — and yet with these foundational rules thoroughly internalised, you will have a solid basis upon which you can start to build your project. If you learn nothing else, memorise the rules and techniques in these opening chapters.

Re-read these rules and techniques on a daily basis — and imagine how you might employ them. Print out these specific pages and put them with your equipment if need be. Gather your equipment and hit the streets. Take these pages with you. Re-read them on the way to your destination and, if it helps, create a best-practices checklist which you methodically work through as you gather footage and experiment with these ideas.

Commit them to memory; utterly internalise them. For quick reference, see chapter seventeen which summarises most of these rules in an easily accessible manner that can easily be used as a quick reference guide in the field.

11. Shots and Compositions Considered

Despite following all of the rules and guidelines outlined in the preceding chapters, it is still possible to shoot an ineffective or poorly composed shot. Too much or too little headroom, or clumsy placement of the audience's focal point, can all have a detrimental effect on the way a shot looks or — more importantly — how it feels.

In *Aftermath: A Portrait of a Nation Divided*, our short film about the 2016 presidential election, there appeared this clumsily framed moment:

Fig. 32. The framing of this shot is of a notably poorer quality than the framing in the rest of the film.

The bodies of the two subjects, relative to the camera, are at a slightly awkward angle. In addition, there is a significant amount of empty, or dead, space around the pair. A more effective way to frame that same shot — or rather, a way the shot could have been improved upon in the post-production process — would have involved the removal of much of this dead space (see Figure 32).

Fig. 33. By zooming in on the footage and reframing the results, a more effective alternative composition reveals itself. This version of the shot was not included in the final cut of the film.

Whilst cropping this shot does not entirely solve the compositional issues at its heart, it does alleviate them. Far more effective than hoping to deal with a problematic image in post-production, however, is paying close attention to one's compositions as they are being constructed, capturing material that does not need to be rescued at a later phase in the production process. Composition is important. Even an untrained onlooker can tell the difference between good and bad composition, even though they may have no idea why one shot feels less satisfying to them than another.

Consider the near-final moments in which the character of Andy emerges from the sewer in Frank Darabont's *The Shawshank Redemption* (1994). As he bursts out of the pipe, the camera tracks with him, following its subject as he moves further from the outlet and into, we might assume, ever-purer waters. He stumbles as he moves, frantically ripping off his shirt. The camera had remained close to Andy throughout most of this process, a reflection of the enclosed space from which he has just escaped. At last, free of his prison-issued clothing, Andy stretches his arms out in jubilation — and the camera cuts. No longer claustrophobically close to its subject, it now looks down upon him, his outstretched arms filling the frame. And then the camera moves, pulling back to free Andy from the metaphorical cell created by the edges of the shot (see Figure 33).

The audience looks down on Andy in his moment of triumph. It is an angle that emphasises his vulnerability in an almost ironic manner. He is vulnerable, to be sure, but this is a shot that is meant to communicate inner strength. It is a brilliant clash of visual and narrative symbols; the triumphalism of the pose versus a camera angle that might otherwise diminish its subject. Even if one were unfamiliar with the rest of the film, the visual language of this sequence alone would serve to communicate its core themes.[1]

Fig. 34. In Frank Darabont's *The Shawshank Redemption*, the triumphant finale sees the camera pan back as it looks down on the protagonist, his arms outstretched. The edge of the frame frequently represents the limits of the observable cinematic universe to the viewer. We know that the subject in the above photograph exists in a space that extends far beyond the limits of this frame — but the edge of the frame, and the subject's relationship to it, nonetheless impacts how an audience respond to the shot. In Darabont's film the frame is not static, as it is in the above homage. The camera movement serves symbolically to free Andy in a way that cannot be replicated in still photography.

Andy's face is never pressed against the edge of the frame during this camera move. An implied degree of looking room exists around his head and face. Had he not been looking up but, instead, was looking straight ahead (and so the audience looking down upon the top of his head, rather than his upturned face), his position in the shot would not have felt as satisfying. As Andy is looking upwards, however, it is the space around the character's face and head that matters — it radiates outwards.

---

1   *The Shawshank Redemption*. Directed by Frank Darabont. Culver City: Columbia Pictures, 1994.

Consider also the way in which the camera movement complements the emotion of the subject's movement. The way the camera spirals away from Andy, as if it were a feather on the wind — free, in other words. This sequence is a masterclass in compositional effectiveness. It does not matter that it comes from a drama. What matters is that it demonstrates how a few seconds of screen time can communicate a vast array of emotions, ideas, and themes through skilled and considered compositional framing.

From a different sort of dramatic movie comes the establishing shot of two comedic, but heroic, robots in *Star Wars: Return of the Jedi* (1983). It is the first time in the film that any of our heroes are seen. R2D2 and C3PO stand in the centre of the frame, walking, with their backs to the camera, down a desert road at the end of which their destination can be seen — the palace of the galactic gangster, Jabba the Hutt.

In this shot, director Richard Marquand uses one-point perspective in order to emphasise the distance the characters must travel; they are on a long and potentially dangerous journey. The shot emphasises the pair's isolation and, with it, their vulnerability. They are dwarfed by virtually every feature around them. In the distance, a huge, alien castle lurches up against the horizon, looming over them. We instinctively understand that this must be the pair's destination. What perils or adventures await them? This shot raises the question; and then primes us for the answer.[2]

One-point perspective effectively conveys distance, allowing for roads and environments to plunge towards infinity. The following shot from *Aftermath* (see Figure 35), though very different in terms of subject and narrative use, works in a similar way to the first shot of R2D2 and C3PO in *Return of the Jedi*.

It is a sunny day in New York. A school bus (a symbol of education, learning, and innocence) disappears down a long road towards an uncertain future. As it turns down the road, a fire engine (a symbol of disaster, danger, and heroism) passes in front of the camera. As the school bus grows smaller, a voice begins to speak about Donald Trump — a controversial topic at the time. In post-production, a slow, subtle zoom was added to the shot, allowing the camera to (virtually) track forward. It thus chases the bus as it moves, albeit far too slow to

---

2   *Star Wars: Return of the Jedi*. Directed by Richard Marquand. Los Angeles: 20th Century Fox, 1983.

11. Shots and Compositions Considered 115

Fig. 35. *Aftermath: A Portrait of a Nation Divided*, directed by Brett Sanders and Darren R. Reid (0:31–0:38).

keep pace with the vehicle (see Figure 35). The movement of the camera emphasises our inability to grasp that which eludes us. As a metaphor for the 2016 election, this was a symbolically effective and relevant shot.

This shot is the result of a combination of factors:

1. Skilful composition on the part of our second unit, who captured this footage.

2. Blind luck, thanks to the unexpected presence of the school bus and fire engine — and a route that took one down a road towards infinity as the other passed in front of the camera.

3. Choices made in the post-production process — the addition of the zoom and the frame's desaturated colour palette.

Despite the way in which all of these factors combined to create a symbolically satisfying shot, it is its composition that serves as the foundation of its success. Even had there been no school bus or fire engine, no desaturation or zoom, the shot would have remained well-composed, containing a degree of inherent beauty.

Infinity and its first cousin, symmetry, are powerful tools. In Jared Hess's film, *Napoleon Dynamite* (2004), there is a moment when the film's protagonist sits perfectly centred on a sofa, with furniture laid out symmetrically at either side. The subject is placed in the dead centre of the frame. The shot encapsulates the perfectly balanced world into which our protagonist fits so uneasily. Despite the fact that even his body is arranged symmetrically, the (literally) slack-jawed subject could not look more out of place.[3]

Symmetrical or centre-framed shots allow filmmakers to use balance in interesting ways — but sparing use of them is encouraged. Life is rarely experienced or perceived in a balanced way and, therefore, a lack of symmetry is to be expected in everyday moments. In Christopher Nolan's *The Dark Knight* (2008), when Christian Bale's Batman confronts Heath Ledger's Joker in the police interrogation room, neither character is centred. There is an inherent imbalance in the scene that reflects the imbalanced nature, not just of the characters, but the nature of their encounter. When watching such scenes, it is helpful to mentally project the 'rule of thirds' grids over them, to see how these guidelines have been followed or disregarded to shape, inform, or subvert a film's core themes.

Note how, in *The Dark Knight*, the tip of Batman's cowl *just* touches the top of the frame in the police interrogation scene (1:25:40–1:30:05). The shot would have felt less clear had the tip of Batman's head, rather than the tip of his costume's ears, been touching the top of the frame. In this case, the details of the character's costume serves to define the precise amount of head room the character requires.

Likewise, the Joker is framed carefully, conforming to the 'rule of thirds', as well as those of head space and looking room. In following those grammatical rules, the Joker is freed to visually demonstrate his disregard for society. To the character's left (our right) there is a small amount of space — not enough to dwarf the character and not so little that the character is pressed up against the edge of the frame. His head has adequate space, allowing the character to exist comfortably within the spatial field defined by the camera. He is technically a prisoner, but he is unconstrained within the frame. Batman, who is much closer to the camera, looms large over his nemesis, the camera looking down slightly upon the Joker, as if to emphasise his vulnerability in the face of Batman.

The camera angle, coupled with the Joker's relative size to the larger-than-frame Batman, signals to the audience that his character *should* be in a vulnerable situation. But, like the shot of Andy's redemption at the

---

3   *Napoleon Dynamite*. Directed by Jared Hess. Hollywood: Paramount Pictures, 2004.

end of *The Shawshank Redemption*, the camera angle is quasi-ironic. The Joker is, of course, where he wants to be; his vulnerability is an illusion, something evident from the clash of symbols (the camera versus Ledger's body language) on display. Even without having watched the film previously, it would be possible to deconstruct the contested power hierarchies at the heart of this scene simply by studying a single frame from it. Such is the power of careful and considered composition.[4]

In the below frames (Figures 36 and 37) from *Aftermath*, we utilised a similar compositional framing technique to that deployed by Nolan and his collaborators. As Figure 37 shows, it follows the 'rule of thirds', but where Nolan's camera looks down towards the Joker, ours is angled up towards our subject, subtly empowering them.

The shallow focus in the shot concentrates the audience's attention onto the subject, encouraging them to pay attention only to their face and, by proxy, the words and signals being issued them: an ironic smile, a nuanced and well considered turn of phrase, a twinkle in the eye. In the context of our film, the environment around this subject was comparatively unimportant, so we were free to shoot with a shallow focus. What mattered was the subject's perspective on Trump and his presidential campaign. By keeping our focus as shallow as possible, the audience was left with no choice but to concentrate their attention entirely onto our subject.

By looking up at the subject, strength is implied. His balanced and reasonable critique of Trump, a man who is, economically speaking, far more powerful than this person, is the source of his strength. As a result, we are reminded that the democratic process can level rich and poor.

The subtle desaturation of this scene (and indeed the entirety of *Aftermath*) helps to provide it with a despondent subtext. The power of the voter is tempered by the possibility of their defeat. In *The Dark Knight*, Nolan does not colour-grade his footage as we do. Instead, he creates a world in which colour is seldom seen but, when it is, it is bright and clear. In this way, the Joker's outfit stands out in a world built (but not graded) around blues and greys. If colour is rare in the world of *The Dark Knight*, it is a deliberate omission by those who inhabit it. They have literally created a world dominated by shades of grey — the contrast between the brightly coloured Joker and the black-costumed superhero

---

[4] *The Dark Knight*. Directed by Christopher Nolan. Burbank: Warner Bros., 2008.

at the heart of the story is sumptuous. Good and evil do battle in a world of moral ambiguity.

With documentary, opportunities to design the colour scheme for an entire world are more limited. But by considering one's compositions and carefully selecting what appears and does not appear within a given frame, strong thematic ideas can still be communicated effectively.

Fig. 36. *Aftermath: A Portrait of a Nation Divided*, directed by Brett Sanders and Darren R. Reid (3:51–4:06).

Fig. 37. *Aftermath: A Portrait of a Nation Divided*, directed by Brett Sanders and Darren R. Reid (3:51–4:06).

12. The Visual Language of Cinema

Film operates much like a language — it has its own grammatical rules and means of construction, much of which you (and your audience) will already understand on a subconscious level. As a result, the audience will have a set of expectations about your work, many of which they will be completely unaware of. Mark Forsyth illustrates the extent of this unconscious expectation thus:

> adjectives in English absolutely have to be in this order: opinion-size-age-shape-colour-origin-material-purpose Noun. So, you can have a lovely little old rectangular green French silver whittling knife. But if you mess with that word order in the slightest, you'll sound like a maniac. It's an odd thing that every English speaker uses that list, but almost none of us could write it out. And as size comes before colour, green great dragons can't exist.[1]

In much the same way, audiences expect films to be constructed in ways they can instinctively understand, utilising conventions and visual cues that trigger emotions and sub-textual understandings. An audience may not be able to articulate the grammatical rules they expect an author to follow, but that will not stop them from being disappointed, or distracted, when these are ignored. Self-aware ironic use and subversion of the rules certainly has its place, but the ability to break them effectively is a rare skill. This chapter summarises some of the medium's most important conventions and grammatical expectations, which you can employ in your own work to communicate, in a purely visual manner, ideas, themes, and subtexts to your audience.

---

1   Mark Forsyth, *The Elements of Eloquence* (London: Icon Books, 2013), p. 39.

Frame Rate

24fps is the frame rate your audience expects. This frame rate is much lower than the human eye is capable of recognising, with emerging mediums, such as video games, regularly employing frame rates of 60fps and above. However, audiences have become so conditioned to expect 24fps in cinematic productions that frame rates other than this can disorientate them, or create the impression of perceived video inferiority. Perhaps the best example of this occurred in 2012 with the release of Peter Jackson's first film in *The Hobbit* trilogy, as discussed in chapter eight. When shooting your own work, aim to shoot at 24fps wherever possible.

Vulnerability, Strength, and Significance through Camera Angles

The relationship between your subject and your camera can be used to communicate important ideas about the subject to your audience. Placing your camera so that it is perpendicular to your subject will create a neutral image perspective, but shooting from a low or high angle can communicate strength or vulnerability. From a low angle, the audience is forced to perceive the subject from a diminutive perspective or, if at a very low angle with the camera close to the ground, from the perspective of a child. As a result, the subject takes on power within the frame, as see in Figure 38.[2]

Conversely, high-angle shots convey vulnerability. By looking down at a subject, the camera emulates physical height, forcing the audience to view the subject from the perspective of an adult or parent.[3] The resultant vulnerability is quickly conveyed to the audience, as seen in Figure 39.

In Orson Welles' *Citizen Kane* (1941), the relationship between characters and their physical surroundings, achieved through careful

---

2   Yoriko Hirose, Alan Kennedy, and Benjamin W. Tatler, 'Perception and Memory Across Viewpoint Changes in Moving Images', *Journal of Vision* 10:4 (2010), 1–19; Andreas M. Baranowski, 'Effect of Camera Angle on Perception Trust and Attractiveness', *Empirical Studies of the Arts* 31:1 (2017), 1–11.
3   Ibid.

framing and positioning of the camera, frequently shapes how the audience relates to the characters. When the eponymous Charles F. Kane delivers his political speeches in *Citizen Kane*, the camera sits at an angle (a Dutch angle), which reflects his increasingly off-kilter world view. Dutch angles involve angling the camera so that the horizon-line of any given shot is no longer horizontal. Dutch angles were used extensively in the live-action *Batman* television show (1966–1968) to depict the similarly off-centre worldview of its villains. .[4] Whilst the 1960s *Batman* show was awash with garish colour palettes, *Citizen Kane* compounded this effect by using shadows to obscure its characters and, thus, their motivations (Batman's deliciously campy villains were never shy about sharing theirs). The position of the camera relative to the subject, and their overall visibility to the audience, were thus able to communicate a significant amount of information to audiences. Rarely are *Citizen Kane* and *Batman* (1966–1968) compared from a filmmaking perspective, but in their use of camera angles at least, they share more in common than one might initially imagine.

There are many ways you can communicate information to your audience by carefully considering the camera's relationship to your subject. By pulling the camera back, the significance of the individual diminishes as they are given less and less on-screen space to occupy. In the above examples, subjects were clearly identifiable. Pulling the camera far enough back, however, can have a devastating impact upon the audience's ability to relate to any person within a frame.[5] Leni Riefenstahl took this to an extreme in *Triumph of the Will* (1935) with wide shots in which all individuality was lost. Masses, not personalities (the Nazi leadership aside), mattered in Riefenstahl's chilling portrait of power and obedience; the significance of the individual rendered utterly meaningless by the power of the collective and their insignificance within the frame (Figure 40).[6]

---

4   It is worth noting that the much more recent Batman-themed television show, *Gotham* (2014–2019) repeats the use of Dutch angles whenever the show portrays Arkham Asylum, in a neat homage to its 1960s predecessor.

5   Sonja Schenk and Ben Long, *The Digital Filmmaking Handbook* (Los Angeles: Foreing Films Publishing, 2017), pp. 219–21.

6   For an insight in Riefenstahl and her Nazi-era films, see Alan Marcus, 'Reappraising Riefenstahl's *Triumph of the Will*', *Film Studies* 4 (2004), 75–86.

Fig. 38. The low-angle shot replicates the perspective of a child looking up at an adult, implying strength in the subject.

Fig. 39. The high-angle shot, which replicates the perspective of an adult looking down upon a child, implies vulnerability.

Fig. 40. From *Triumph of the Will* (1935), directed by Leni Riefenstahl (1:02:55–1:08:02).

Wide Shots, Close-Ups, Mid-Shots

Welles and Riefenstahl both demonstrate the power of the wide shot. Riefenstahl used them to obliterate individuality and to create a sense of vast scale. In Welles's hands, they emphasise individuality through careful, precise compositional placement. Typically, however, wide shots are more functional in nature, serving primarily to establish physical context. A film that takes place in New York, for example, would benefit from wide shots that show the city's iconic skyline. Such shots serve to establish a spatial context for an audience and are therefore an important part of most productions. In terms of communicating the thoughts and emotions of a subject, however, the mid-shot and the close-up are of particular importance to the filmmaker-scholar.

Fig. 41. A close-up will allow your audience to read subtle facial expressions and micro gestures not otherwise evident in mid-shots (and certainly not in wide shots).

A mid-shot (typically encompassing a subject from at least the top of their head down to their lower abdomen) helps to provide a broad overview of a person's body language. Conversely, a close-up (which focuses almost all attention on the subject's face and/or eyes) helps to reveal a person's emotional state by laying bare otherwise imperceptible changes in their facial expressions. The mere act of cutting to a close-up tells the audience that they need to begin paying greater attention to the subject's internal emotional state — often expressed through their eyes. In a documentary, a subject might talk directly to the camera but a cut from a mid-shot to a close-up would focus attention on the emotional dimension of their discourse.⁷ This is helpful in moments of candour or complete vulnerability.

This requires forethought on the part of the filmmaker-scholar, however. Before an interview is conducted, they must anticipate if/when their camera should move closer to their subject. In some instances, this may require running more than one camera at a time; alternatively, filmmakers can ask their subject to repeat an answer, adjusting the camera setup as necessary between takes. These three shots (wide, mid, close) each serve a different intellectual purpose. Wide shots are about context (or placing a subject in context). Mid-shots provide detail about a subject, allowing audiences to read their body language. Close-ups are about connecting an audience with a subject on a deeper, more emotional level. If the mid-shot is about body language, the close-up is about micro gestures. Once your camera is set up and recording footage, remain aware of the type of shot you are recording, weighing it against the content you are capturing. If you are engaged in an interview and the discussion becomes more personal or emotional, it may be appropriate to switch from a mid-shot to a close-up.

Aspect Ratios

Fig. 42. The standard 16:9 aspect ratio will fill the entirety of a modern widescreen television.

---

7   Mercado, *The Filmmaker's Eye*, 29–70.

Aspect ratios can have a powerful impact on how we interpret what we see on screen. Often unnoticed by audiences, aspect ratios (and changes between them) can serve as powerful visual cues. The 4:3 Academy ratio, for instance, is most closely associated with films from the golden era of Hollywood and its use can evoke a feeling of nostalgia. In *The Grand Budapest Hotel* (2014), director Wes Anderson cuts between the modern 16:9 (widescreen) aspect ratio for scenes set in the current day, and the 4:3 aspect for scenes that occurred in the 1930s. This subtle change likely went unnoticed by most members of the audience, but nonetheless served to signal important information to them.

As most modern cameras capture footage in the 16:9 aspect ratio (which fills a standard widescreen television), this is the ratio that feels most comfortable for documentary footage. Most documentarians do not alter their aspect ratio; as a result, audiences have come to expect such films to be presented in 16:9. However, the use of, for example, the 4:3 ratio may be viable should the filmmaker-scholar wish to evoke the period in which this was the standard cinema ratio. In addition, the use of the more cinematic 21:9 aspect ratio may be appropriate when the filmmaker-scholar wishes to evoke the feeling of modern cinema. This ratio creates a narrower field of view and is a common feature of modern content creation. Using such an aspect ratio for the entirety of a documentary may, however, prove distracting to audiences. Just as the 4:3 aspect ratio is closely associated with media from the first half of the twentieth century, the 21:9 aspect ratio is closely associated with drama and big-budget blockbusters. The 16:9 ratio, in contrast, is the ratio that feels most familiar to viewers of documentary content.[8]

Most cameras will shoot only in the 16:9 aspect ratio. In order to accomplish a 4:3 or 21:9 look, it will be necessary to frame shots with these aspect ratios in mind. Strips of card can be attached to the digital display on one's camera (being very careful not to cause permanent damage to your device) to create a 4:3- or 21:9-proportioned viewfinder. This will allow the camera operator to compose shots suitable for these aspect ratios. The camera will still capture standard 16:9 footage, but the addition of simple black bars (along the top of one's footage, or down the side) in post-production will produce a fair approximation of the desired aspect ratio.

---

8   Harper Cossar, 'The Shape of New Media: Aspect Ratios, and Digitextuality', *Journal of Film and Video* 61:4 (2009), 3–16.

Fig. 43. The 4:3 aspect ratio tends to evoke the era of early Hollywood. This aspect ratio is useful for generating a sense of nostalgia.

Fig. 44. A 21:9 aspect ratio is common in modern cinema. This aspect ratio is useful in evoking the sense of hyper-reality that so often accompanies modern films.

13. Interviews

Fig. 45. Watch the video lesson on conducting interviews. http://hdl.handle.net/20.500.12434/47ac0bf7

Interviews are often at the heart of documentaries. They will provide you with an opportunity to engage with other scholars, or to create new primary artefacts based upon the lived experiences of participants, activists, and witnesses. Conducting a successful interview involves balancing a number of factors, from ethics and safety, to intellectual preparation and writing the questionnaire.

Conducting primary interview research for your documentary project will add depth to its analysis. Whilst questionnaire data can be deployed in the narration or as statistics on screen, filmed interviews are an excellent addition to a documentary and provide both depth and production value. It is in these interviews that we experience the tension between ideas and perspectives, and the evocation of life stories. In fact, when we think about documentary films, one of their most important and visible features is frequently the interview.

Until fairly recently, conducting professional-style documentary interviews has been somewhat out of reach. Access to suitable equipment was often limited by its expense and transportability. However, with the democratisation of filmmaking technologies, filmed interviews have become increasingly viable, especially with advances in online video-calling.[1] The very rudiments of the humanist's study — a written record on paper — is also undergoing radical change. As our means of communication and documentation evolve, so too does the framework in which they may be studied and articulated. Borrowing from the methods of oral historians, you can use interviews in your own research, producing primary data as well as developing archives of their subject's lived experiences. This chapter will provide a theoretical discussion about the application of oral history methods, as well as providing a step-by-step guide to interviewing, designing questions, the ethics of interviewing, the role of the interviewer, and the limits of interview data for academic use.

Oral History and Interviewing

By borrowing from the oral historian, filmmaker-scholars can produce their own primary materials. Whilst scholarship in the humanities is historically rooted in the analysis of written materials from state archives and newspapers, for example, and published in the same form, oral historians operate beyond these parameters, gathering novel interview material as the basis of their work. In the same way that oral historians' innovations in historical method added to the record by providing a voice to those often denied visibility in traditional archives, the filmmaker-scholar has the capacity to platform these voices. The digital revolution has fostered an academic environment wherein the analytical skills of the humanist can be readily captured by new technologies and disseminated by new and emerging distribution channels.

---

1 Oral History Society (nd), *Getting Started*, https://www.ohs.org.uk/advice/getting-started/3/#:~:text=Be%20able%20to%20record%20uncompressed,use%20different%20types%20of%20card); L. Abrams, *Oral History Theory* (Abingdon: Routledge, 2016), p. 82.

By moving to generate their own primary material, the pioneers of oral history in the 1960s and 1970s opened up the study of the past to include groups often omitted from the archival record.[2] As Robert Perks and Alistair Thomson put it:

> While interviews with members of social and political elites have complemented existing documentary sources, the most distinctive contribution of oral history has been to include within the historical record the experiences and perspectives of groups of people who might otherwise have been "hidden from history", perhaps written about by social observers or in official documents, but only rarely preserved in personal papers or scraps of autobiographical writing.[3]

The harnessing of the availability of sound-recording technologies was so profound a shift in the way that the historical record could be expanded that Arthur Marwick called it a 'mini-Renaissance'.[4] The drive to uncover submerged layers of the past has seen 'the experiences of a number of groups who had traditionally been disregarded by conventional histories: women, gays and lesbians, minority ethnic groups and the physically and learning disabled' become important aspects of the record.[5] The addition of a visual element, capturing nuances of body language and inflection, can only deepen the potential of this method. The Oral History Society breaks the advantages of this approach down into four key elements:

- A living history of everyone's unique life experiences.

- An opportunity for those people who have been 'hidden from history' to have their voice heard.

- A rare chance to talk about and record history face-to-face.

- A source of new insights and perspectives that may challenge our view of the past.[6]

---

2   Simon Gunn and Lucy Faire, *Research Methods for History* (Edinburgh: Edinburgh University Press, 2011), p. 18.
3   Robert Perks and Alistair Thomson *The Oral History Reader* (London and New York: Routledge, 1998), p. ix.
4   Arthur Marwick, *The Sixties* (Oxford: Oxford University Press, 2011).
5   Lynn Abrams, *Oral History Theory* (London and New York: Routledge, 2016), p. 4.
6   Oral History Society, https://www.ohs.org.uk/

Oral historians choose their interview subject and shape the contours of that encounter; they are the 'only historians who deal exclusively with the living'.[7] In addition, direct encounters with one's subjects can create new opportunities to gather other forms of evidence, with interview subjects often being in a position to provide further written documents, photographs, and other research materials, which might not otherwise have been available. As a consequence, the 'confines of the scholar's world are no longer the well-thumbed volumes of the old catalogue. Oral historians can think now as if they themselves were publishers: imagine what evidence is needed, seek it out and capture it'.[8] By embracing the interview as the means to reconstruct the past or present, oral historians have significantly widened the source-base upon which we can draw. . If we position the documentary-making humanist as a publisher in a trans-media environment, that widening becomes even more apparent. Not only do they collect and store data, stories, and perspective, they now actively disseminate those accounts in a way that captures the nuance of body language and facial expression, as well as changes in tone, delivery, and emphasis.

Designing an Interview

When planning for your interview there are four main approaches that might be taken: structured, semi-structured, unstructured, and focus groups:[9]

Structured interview: This is the most rigid form of interview, in which you arrive at the interview with a pre-determined set of questions. You will only ask these questions. Structured interviews are useful if, for example, you have multiple interviews planned and you wish to offer a uniform experience for your interview subjects. This adds consistency and, perhaps, a way to ensure that you can compare and contrast views in your documentary. In many ways, this style of interviewing is like an oral questionnaire.

---

7   Donald A. Ritchie, *Doing Oral History* (Oxford: Oxford University Press, 2015), p. xiv.
8   Paul Thomson, *The Voice of the Past: Oral History* (Oxford: Oxford University Press, 1988), p. 28.
9   Patrick McNeil and Steve Chapman, *Research Methods* (London: Routledge, 2005), p. 56.

Semi-structured interview: Like a structured interview, this approach also requires a pre-planned questionnaire. However, rather than being entirely pre-determined, a semi-structured interview provides the flexibility to ask follow-up questions. It requires you to design and plan the exchange, but it eschews the rigidity of a fully structured process; it is not essential that each question is asked, nor that your interviews all follow the same sequence. This is likely to be the style of interview that documentary makers will find the most useful — it ensures that the key areas of the project are covered but also allows for flexibility. A semi-structured interview would allow the interviewer to adjust their questions in response to the answers given, enabling them to elicit the best responses from each subject. Together with this flexibility, this approach retains an overall structure, ensuring that common themes and issues are covered by all of your different interview subjects.

Unstructured interview: This type of interview requires less formal planning (though not less preparation). Although the broad parameters of the exchange will be understood in advance, no formal questionnaire would be utilised, relying instead upon the interviewer's familiarity with the topic or their chemistry with the subject. Such encounters may provide unexpected results that might not have emerged from a more rigid line of questioning. However, what is gained by limited planning is potentially lost if the resultant discussion fails to engage with core ideas or themes — issues can easily be forgotten in the moment, and important issues left unexplored. Unstructured interviews are most appropriate in a spontaneous context, such as during a protest or emergency when circumstances do not allow for any advanced planning.

Focus groups: This a group interview. The interviewer acts as mediator or chair of a panel-style discussion about a given topic. It is a useful method if there are a large number of available interview subjects or, for example, there is an opportunity to interview a whole department of an organisation. Focus groups might draw out debates between participants and necessitate not only listening skills but also mediation, ensuring that dominant voices are controlled and quieter ones encouraged. Focus groups also lend themselves to longitudinal studies whereby repeat interviews can eke out changing (or static) attitudes.

Formulating Interview Questions

In designing the interview, the phrasing of questions is very important. Different types of questions lead to different types of responses and, of course, the questions must be designed to avoid leading the interview subject towards a pre-determined response. A list of twenty-five questions should be drawn up for a sixty-minute encounter.[10] This might be broken down into five key areas, each comprising five questions per section. In other words, the interview starts with a general question before becoming more focused. Donald Ritchie has argued that a two-sentence format is preferable, whereby the first offers the problem, and the second poses the question.[11] This is sometimes referred to as 'funnel interviewing'.[12] There are, of course, many ways of phrasing questions; this will determine the nature of the response you wish to capture: do you want single-word answers or longer, more considered, discussion?

When you are designing your interview questions, there are two main types of questions that you might pose your interviewee — open and closed questions. Open questions invite longer, more involved answers. Closed questions tend to elicit short, decisive answers. You will no doubt want to include a mixture of open and closed questions, but you will need to plan the order in which you pose them to your subject.

In general, it is best to start with open questions; allow your subject to ease into the topic and express their thoughts. As you progress through the questions for each section of the interview, you can start to round each discussion off with a closed question. For example, in a discussion about the history of silent film, you might ask your interviewee:

> To what extent was Charlie Chaplin the master of the silent film era?

This is an open question: rather than inviting a 'yes' or 'no' answer, it invites a longer and more considered response which will provide much deeper insight and consideration. These are sometimes also referred to as dialogical questions, as they encourage reflection and the creation of an extended discourse.[13] Such a question would likely provide much

---

10 Thomson, *The Voice of the Past*, pp. 225–26.
11 Ritchie, *Doing Oral History*, p. 81.
12 Ibid.
13 Higher Education Academy (n.a.), *Historical Insights Focus on Research: Oral History* (Coventry: Warwick University Press, 2010), p. 28.

deeper material for a documentary than its closed equivalent. In purely practical terms, this would provide you with significantly more material on which you can draw during the editing process. It would also allow you to compare and contrast the responses of different interviewees.

In contrast, when discussing the significance of Chaplin's filmmaking, it might be interesting to evoke a definite answer about the quality of his work. Asking a closed question would encourage this. For example, you might ask:

> Did Charlie Chaplin make the best silent films?

This question invites a 'yes' or 'no' answer; your interviewee will either agree with the proposition or not. Closed questions are appropriate if you want a definitive answer to specific question. They are also useful as a final summation of a topic, perhaps to distil a conversation down to a final conclusion.

There are other types of questions, such as anchoring questions that ask the subject to place themselves at a particular point in time. So, for example, you might ask:

> Where were you when you saw Charlie Chaplin's *Limelight*?

This question invites the interviewee to reveal a date, place, and time. It also helps to indicate the interviewee's age and elicit some of their socialisation.

The Phrasing of Questions

Closed questions:
'Did you....'
'Do you think that....'
'Do you agree that....'

Open questions:
'To what extent....'
'In what ways....'
'Tell me about....'

The Role of the Interviewer

As well as developing certain research skills, filmmaker-scholars must also learn to be effective interviewers. It is essential that interviewers develop a new set of skills that include an understanding of human relationships.[14] Having framed the contours of the encounter in the research documentation (discussed below), the interview should settle into a rhythm within the first twenty minutes. Fundamental to the interview is that, like an oral historian, the interviewer 'has to be a good listener, the informant an active helper'.[15] Indeed, patience and considered prompts following natural pauses in the conversation will keep the dialogue going: do not interrupt the subject, only follow with additional questions once they have finished. According to The Higher Education Academy's oral history guide, interviewers should:

- **Show interest:** by active listening, looking interested (nodding and smiling rather than making verbal sounds of appreciation), picking up on what has been said when it is appropriate and in natural breaks in the conversation.

- **Maintain eye contact:** although beware that this is subject to cultural contexts.

- **Reassure:** that what is being said is interesting, even when it might not seem so; it is surprising how often what seems to be mundane turns out to have significance when it is subsequently analysed.

- **Empathise when appropriate**: be compassionate, but try to avoid empathising with experiences that are simply outside of the interviewer's knowledge or experience.

- **Avoid making assumptions:** try to ask questions to test assumptions. If information seems ambiguous, find ways of asking for clarification.

- **Avoid disagreeing or arguing:** interviewees can have values and beliefs that are at odds with those of the interviewer, but the session is about the interviewee's life, including their ideological orientations. It is not about the interviewer's prejudices, assumptions, and beliefs (no matter how well-intentioned they might be).

- **Be relaxed and measured**: avoid hurrying through the interview and skipping from topic to topic — think about the

---

14 Thomson, *The Voice of the Past*, p. 30.
15 Ibid, p. 31.

interview flow and keep questions and prompts short and clear.

- **Use emotional intelligence:** to connect to the interviewee and fine-tune when and how questions should be asked.[16]

The Interviewer/Subject Relationship

The interview process is, by definition, an active one, whereby the communication between the two actors must develop what has been called a 'conversational narrative: conversational because of the relationship of interviewer and interviewee, and narrative because of the form of exposition—the telling of a tale'.[17] In that sense, then, we must, as scholars conducting interviews, and thus the creators of new primary material, acknowledge that we are involved in the creation of artefacts, unlike our peers who rely on archival material alone. We must, therefore, carefully consider our role — the impact of our own subjectivities — in the production of the primary data derived from that process.

The active participation of the interviewer in this 'conversational narrative' disrupts their attempts at neutrality as they fundamentally help to shape the story. In other words, the memories, experiences, and reflections elicited by the interview process are not an objective truth about the past; they are creative narratives shaped in part by the personal relationship that facilitates the telling.[18] This methodological conundrum has been referred to as intersubjectivity, a phenomenon that 'describes the interaction — the collision, if you will — between the two subjectivities of interviewer and interviewee. More than that, it describes the way in which the subjectivity of each is shaped by the encounter with the other'.[19] For many scholars this creates a validity problem, which may prompt some to question or even refute data that is collected in this way.

In addition to the perceived issue of intersubjectivity, and the active participation of the researcher in shaping the historic record, others

---

16   Higher Education Academy, *Oral History*, p. 31–31.
17   Perks and Thomson, *The Oral History Reader*, p. 44.
18   Abrams, *Oral History Theory*, p. 58.
19   Ibid.

have noted the potential rift between truth and memory. Indeed, the filmmaker-scholar, like the historian,

> asks people questions to discover four things: what happened, how they felt about it, how they recall it, and what wider public memory they draw upon. At the heart of this lies memory. Memory and the process of remembering are central to oral history. The recollections of memory are our primary evidence just as the medieval manuscript or the cabinet-office minutes are for historians working within other traditions[.][20]

Indeed, this idea lies at the heart of A. J. P Taylor's often used[21] but uncited disapproval of oral history as 'old men drooling about their youth' — a scathing commentary on the ability of interviews to generate objective recollections given the fallibility of human memory, and the propensity of such recollections, unlike written documents, to change over time.[22] This does, however, seem to ignore the fact that written testimonies or minuted records are likewise based on the selection of information committed to paper, or the memories of those, for example, writing their memoirs. It also ignores stark discrepancies between different ethnic groups, genders, social classes, and sexualities within the archive.

So, whilst '[d]ealing with memory is a risky business',[23] it is the fundamental ingredient of a documentary film's ability to engage a wide range of voices. In addition, providing that the interview is constructed in a way that avoids leading the interviewee, it unlikely that the interviewer can subvert the historic record as '[p]eople remember what they think is important, not necessarily what the interviewer thinks is most consequential'.[24] In that sense, the objective is 'searching not for fact, but the truth behind the fact'.[25] Oral historians have helped us to understand the distinctive qualities of recorded memory.[26] Indeed,

---

20 Ibid, p. 78.
21 This quote first appeared in Brian Harrison's 'Oral history and recent political history', *Oral History* 1 (1972), 30–48, and is likely derived from personal correspondence rather than Taylor's published writings.
22 Ritchie, *Doing Oral History*, p. 10.
23 Ibid, p. 15.
24 Ibid.
25 Ronald J. Grele, *Envelopes of Sound: The Art of Oral History. Second Edition* (1985; New York: Greenwood Publishing, 1991), p. 129.
26 Simon Gunn and Lucy Faire (eds), *Research Methods for History* (Edinburgh: Edinburgh University Press, 2011), p. 102, ch. 7.

whilst the humanist (and historian) usually relies on archival sources, the 'use of interviews as a source for professional historians is long-standing and perfectly compatible with scholarly standard'.[27]

The Ethics of Interviewing

Before interviews can be arranged and filmed, there are some important steps that must be taken. These ensure your safety as an interviewer and that of your subject. It is 'essential that interviewees should have confidence and trust in interviewers, and that recordings should be available for research and other uses within a legal and ethical framework which protects the interests of interviewees'.[28] Most universities and institutions will have their own ethics procedures to ensure the safety and well-being of the researcher and participants. It is absolutely essential that these are followed, both from a legal and moral perspective. In particular, and applying the methods of the oral historian, the interview process has the potential to be an emotive experience whereby, depending on the topic, the participant may be speaking about troubling aspects of their life. Indeed, during the interview process, the participant 'may breach a lifelong silence or make new sense of experience, and perhaps find recognition or even catharsis through stories that have never been easily told. At worst, if the dialogue opens wounds that are still raw and offers no way to make new, affirming meaning, it risks a "dis-composure" of safe stories and settled identities"'.[29] In order to safely navigate this process, there are a number of key steps that must be taken.

As a starting point, you must produce two documents that you can send to your interviewee in advance of encounter. The first is a Participant Information Sheet; this document describes your project's aims, objectives, and scope. As part of this, it is important to explain why you have asked the participant to be involved, what the participation (i.e., the interview) involves, how you will use and store the footage, and the contact details of a person who can handle any complaints they

---

27 Thomson, *The Voice of the Past*, p. 26.
28 Oral History Society, *Is Your Oral History Legal and Ethical?* https://www.ohs.org.uk/advice/ethical-and-legal/
29 Gunn and Faire, *Research Methods for History*, p. 108.

may have once the interview has been concluded. The second document is an Informed Consent Form. This asks the participant to sign off on the aspects of the exchange that they are happy with. These will take the form of declaratory statements which ask, for example, whether they are happy to be named or for you to use their footage in your documentary film.

The Interview Process

1. Make a list of people you would like to interview for your documentary film.

2. Conduct your preliminary research to gather contact details of your potential interviewees.

3. Contact your list of interviewees either by telephone, email, or via social media with a short outline of your research and why you have contacted them. Avoid using the word 'interview' as this can sound overly formal. Instead, ask whether they would be willing to have a 'chat' or 'conversation' about your topic.

4. Once they have provisionally agreed to take part, forward your Participation Information Sheet and Informed Consent Form to ensure that they know what taking part involves.

5. Arrange the date, time, and location of the interview.

6. The interview should take place in a safe space, mutually agreed, and in a room without distractions such as televisions and telephones.

7. On the day of the interview, set up your equipment and build some rapport with your subject as you position them and the equipment. Consider the rule of thirds (see chapter ten) when framing the interview subject. The interview sections of your documentary film are as important to your visual grammar as any other aspect of your project.

8. Before you start the interview, make sure your subject introduces themselves to the camera, providing their name, the purpose of the interview, and their consent to being filmed.

9. The interview should last no longer than sixty minutes.

10. Ask one question at a time — be clear in your questioning.
11. Start with open questions that are broader before moving to more incisive questions; conclude with closed questions to draw out more definite answers.
12. Make eye contact as your subject answers your questions — listen intently and provide a relaxed environment.
13. Do not interrupt the response; wait for a natural pause before moving on or asking a follow-up question.
14. Do not be combative or argue with your interviewee.
15. Allow your subject to speak 'off the record' if they wish.
16. Following the final question and response, ask if they have anything else to add or whether they have any questions.
17. Thank your subject for taking part.
18. Ask them to sign the Informed Consent Form.

Sample forms and templates (Participant Information Sheet, Informed Consent Form) are included on the following pages.

Participant Information Sheet Template

[Title]

[Short paragraph of your documentary's key aims]

What is the purpose of the study?

Why have you been chosen?

What will participation involve?

You should know that:
- The interview will take place at an agreed location that ensures the safety of both interviewee and interviewer.
- The interview will be recorded, with your consent.
- Initially, access to the interview recording will be limited to [name] and academic colleagues and researchers with whom [he/she] might collaborate as part of the research process.
- Both summaries of, and direct quotations taken from, our conversation, attributed to yourself by name, will be used in a documentary film and academic publications unless you wish these comments to be anonymised. If you wish parts of the interview to be regarded as 'off the record', please indicate that this is the case.
- The actual footage will be stored on [insert storage solution].

Do I have to take part?

What will happen to the results of the study?

Who should you contact for further information?

If you wish to seek further information or have a complaint about the researcher, please contact:

Researcher:

Name:

Job Title:

Address:

Email:

Telephone:

Director of Research

Name:

Job Title:

Address:

Email:

Telephone:

Informed Consent Form Template

[Project Title]

[Short paragraph of your documentary's key aims]

Before you decide to take part, it is important for you to read the accompanying Participant Information Sheet.
If you have any questions or queries about the interview, please contact the researcher using the details listed below:

Name:
Job Title:
Address:
Email:
Telephone:

By signing this form, I agree that:

	Please initial
1. I confirm that I have read and understood the Participant Information Sheet for the above study and have had the opportunity to ask questions.	☐
2. I understand that my participation is voluntary and that I am free to withdraw at any time without giving a reason.	☐
3. I agree that this interview may be recorded and stored electronically.	☐
4. I understand that, unless I indicate otherwise, the interviewer may reproduce material gathered from this interview as attributed quotations in their documentary project, and subsequent academic publications.	☐
5. I understand that if I wish any part of this interview to remain in confidence, this is possible, and I should indicate to the interviewer which passages should be treated as 'off the record'.	☐

6. I do not expect to receive any benefit or payment for my participation.	☐
7. I agree to take part in the research project.	☐

Participant(s) Details:

Name of participant(s):

Signature(s) of participant(s):

Date:

Name of Researcher:
Address:
Email:
Telephone:

Signature of researcher:
Date:

14. Recording Audio and Creating Soundscapes

Your audience requires clear and well-recorded audio. They might be willing to accept poor imagery, but sound quality and, most importantly, clarity is non-negotiable — the sound-track matters.[1] Your audience will immediately be reminded that they are watching a film if they have strain in order to hear its dialogue, and, in so doing, their immersion will be broken. Your audience needs to be able to invest their intellectual energy into what your film is saying, not squander it as they struggle to discern individual voices.

The debut of material from Christopher Nolan's *The Dark Knight Rises* (2011) was slammed by audiences and critics for precisely this reason. In a preview of the film's opening sequence, Tom Hardy's Bane, the film's central antagonist, spoke in a voice that was muffled and difficult to hear. Nolan, a highly skilled filmmaker, had sound reasons for muffling Bane's vocals — the character wears a mask and, as such, his voice should have been difficult to hear. Audiences, however, were utterly unwilling to accept real-world logic in a cinematic presentation. Whether realistic or not, audiences demand clear audio in their films.[2]

No one wants to sit through a film in which the dialogue is not clear or easily understood. As a result, Bane's voice was made to boom in the final mix of *The Dark Knight Rises*. It dominates much of the film and rarely is it difficult to hear or understand.[3] Never cut corners on audio

---

1 Barry Callaghan, *Film-making* (London: Thames and Hudson, 1973), pp. 88–103.
2 Borys Kit, '"The Dark Knight Rises" Faces Big Problem: Audiences Can't Understand Villain', *Hollywood Reporter*, 20 December 2011, https://www.hollywoodreporter.com/heat-vision/dark-knight-rises-christian-bale-batman-tom-hardy-bane-275489
3 *The Dark Knight Rises*. Directed by Christopher Nolan. Burbank: Warner Bros., 2008.

quality. Wonderful sequences can be ruined and made unusable by poor or inaudible sound.

Recording Sound on Site

Recording clear audio essentially comes down to two factors — your recording equipment (i.e., the presence of a quality microphone and sound recorder) and its proximity to the sound you wish to record. By default, you will record some sound as you work in the field since practically all cameras have a built-in microphone, but you should work on the assumption that the quality of audio captured with such a device is likely to be poor, if not outright unusable.

On a DSLR, for instance, if you adjust the focus of your shot, the internal microphone on that camera will likely pick up the mechanical sound of your focus mechanism. As a result, if you have a subject speaking on camera, their dialogue will probably be buried under the loud, unpleasant sound of shifting and grinding gears. To make matters worse, sounds closer to your camera's internal microphone will be much louder than more distant sounds. When shooting an interview, the internal microphone in your camera will pick up the ambient noise around it far more effectively than it captures the voice of your subject. Instead of relying upon your camera's internal microphone, you should instead utilise other audio devices and microphones to ensure you capture clear, usable audio. This separately recorded audio track can be added to your footage during the post-production process.

On-site sound can be captured in several ways.

Rough and Ready

Place a mobile phone, recording via a sound-recording app, near to the person speaking. A mobile phone attached to a pole and held over the person's head, but out of shot, will capture relatively poor-quality audio — but it will still be better than the audio captured natively on a camera's built-in microphone.

Lavaliere Microphones

To record a person speaking, they should ideally be given a lavaliere microphone. These are small microphones that can be attached to the lapel of a person's jacket. Although there are very cheap models available, we would recommend that you do not start at the lowest possible price. Such devices tend to capture muffled, poor-quality sound. We have had excellent experiences working with sound equipment by Rode. The basic Rode lavaliere microphone costs approximately $60 but captures a clean sound profile which works perfectly well for on-site discussions in documentary films.

Such microphones usually also require a sound recorder; however, some Rode microphones can record directly onto your smartphone. Using this solution, you will not require a standalone sound recorder, reducing your overall equipment cost.

Run and Gun

A lavaliere microphone is ideal for recording interviews, but it is not an ideal solution for recording more ad-hoc material. In situations where you cannot spend time wiring your subject for sound, you can usually add an external microphone to your camera. This will give you the option of recording higher-quality ambience and, if you purchase a directional microphone, the opportunity to capture audio emanating from a specific direction. Directional microphones pick up more of their sound profile from the direction in which they are pointed, allowing you to 'run and gun' with your camera/microphone setup. For events that are unfolding quickly, this solution will allow you to capture usable sound that will not require you to wire up your subjects with lavaliere microphones. As with capturing video, capturing usable sound becomes easier with practice and experience. From an early stage, filmmakers should experiment to ensure they identify the solution that will work best for them.

Clipping

When recording sound, you should pay attention to the amount of audio being picked up by your microphone/recording device. If a microphone is too far away from your subject, sounds may be inaudible or unclear. If it is too close, however, more sound may be entering the microphone than the device can handle. This is called clipping, and it creates a nasty, distorted sound which you should aim to avoid. The result is a sound which cannot be removed in post-production.

It is possible to visually identify clipping. If too much sound enters the microphone it will stop recording sound data at both the lower and upper extremity of the device's range. On an audio recorder, recorded sound should look something like the sound wave seen in Figure 46a. Both the upper and the lower end of the sound wave are within the upper and lower limits of the recordable field — this is usable sound. Clipping, on the other hand, looks like that seen in Figure 46b.

Note how the sound wave hits both the top and bottom of the above field. The sound information that would appear above and below these sections simply does not exist, so rather than a smooth, curved sound wave, clipped areas instead end abruptly.

In order to avoid this, always test your microphone and recording environment prior to recording. Attach your lavaliere microphone and speak at the volume you intend to record (or have your subject do the same). If the sound wave is very small, you should probably move the microphone closer to your subject's mouth. If the sound wave is too large and clipping occurs, or it looks like this might occur, move it further away or reduce the amount of sound your device is attempting to record.

On-Site Tips

When you activate your microphone/sound recorder, look at how much background ambience is being picked up by the recorder. If there is a sound wave of significant size already, you might struggle to hear the person being recorded unless the microphone is placed close to their mouth. This, however, can increase the risk of clipping.

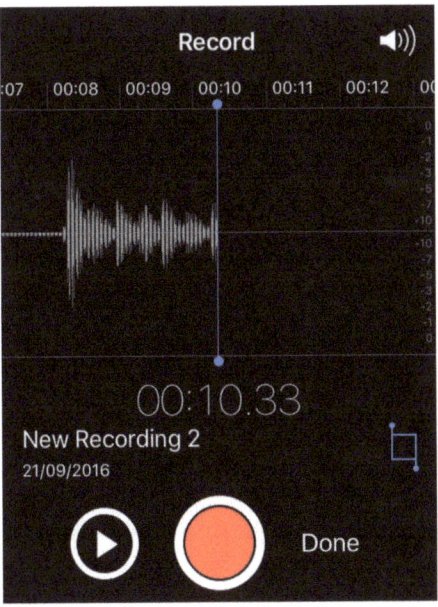

Fig. 46a. The sound wave fits comfortably within the recordable field.

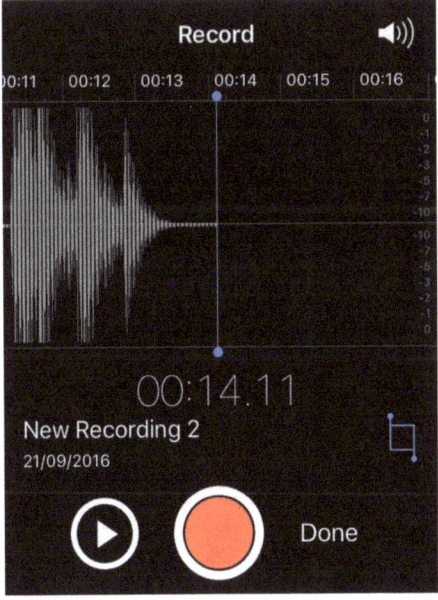

Fig. 46b. The device's recording sensitivity is too high, or the microphone is too close to a sound source.

If this happens you will either need to wait until there is less ambient sound or move to a quieter location. You should not be afraid to change your location if capturing clean audio is likely to prove difficult or impossible. As beautiful as a given setting may be, if you cannot record usable sound, the footage will be useless. Remember, when recording sound, particularly in the field, you must consider both the audio and the visual elements you will capture. As a result, you should reconsider locations such as busy cafes, particularly if the level of noise produced by the clientele is consistently loud or prone to unpredictable spikes. It only takes one person with a booming voice to turn a beautifully filmed section into an unusable piece of footage.

To that end, prepare contingency plans if you are planning on recording audio on site. Plan A should focus on shooting in your preferred location, but if there is an unpredictable noise profile, an alternative location will be needed. Your contingency should therefore be a location where you have much greater control over the ambience.

In the worst-case scenario, you can record new audio over pre-existing footage in post-production, having a subject repeat what they said in a more controlled environment. Syncing up such audio is, however, tedious and difficult to accomplish. You will have to line up the new audio very closely with the recorded footage; even a small discrepancy between sound and visual elements can pull an audience out of the moment. Instead, your priority should be on recording usable audio on site in the first instance.

Engineering Ambience

Film is often described as a visual medium, and there is a lot of truth to that idea — but it is not the whole truth. Sound, its presence or absence, is a huge part of the cinematic experience, even if it is not always the most important aspect. Although most of the information communicated via film is transmitted visually, an appropriate and enriching soundscape is important. Even in the silent era, sound was an important part of the process. Live musicians and orchestras — and sometimes sound effects — accompanied 'silent' films.[4]

---

4   Rick Altman, *Silent Film Sound* (New York: Columbia University Press, 2004).

As already discussed, recording clear vocal audio is essential — but so too is ensuring an appropriate ambience for your chosen visuals. The poor quality of most built-in microphones, coupled with their tendency to pick up nearby sounds (often the sounds made by the camera or its operator), can create serious issues.

Ambience can be recorded on site or it can be sourced from a sound-effects archive and added to your footage during post-production. For many, recording authentic ambience is important — but the internal microphone in most digital cameras will struggle to capture a balanced or usable ambience. Instead, connect a sound recorder or smartphone to a multi-purpose microphone to capture a space's ambient sounds. As with recording footage, capture more audio than you require, and beware of objects or people near to your microphone, as any noises they make will feature prominently in your recorded ambience.

Just as when shooting in low-light situations, do the best you can with the equipment you have to hand. If you do not have a dedicated sound recorder or external microphone, record local ambiences with whatever equipment is available to you. The resultant audio may prove unsuitable or unusable, but if the conditions are correct, and if your luck holds out, you may record some usable ambience. If this is not successful, it is possible to engineer ambience during post-production. A wider range of sound-effect archives can be found online, where different ambiences can be purchased or downloaded freely. Applying these soundscapes to your existing footage is not difficult, though some sounds may need to be layered, depending on what is happening in your footage (see chapter twenty-three).[5]

Ambient sounds rarely need to be synced up to the original video; they provide atmosphere, not detail. If specific events occur on screen, however, such as a person in the foreground coughing, the appropriate audio, which can also be sourced from a sound-effect archive, can be easily applied at the correct moment. There are numerous factors that can prevent you from using the audio you capture in the field. Blowing wind can wreak havoc with poor quality, unshielded microphones, whilst off-camera activities can create recorded audio that does not feel appropriate for the shots you have captured. In such instances, employing a pre-recorded ambience may be a necessity.

---

5   Roey Izhaki *Mixing Audio: Concepts, Practices, and Tools* (Burlington: Focal Press, 2013), pp. 5–11.

There is a wide range of sound-effect archives online, some offering paid products, others offering free downloads. When using these archives, attention should be paid to ensure that you select an ambience that matches the visual element of your film. Like recorded dialogue, an audience is unlikely to appreciate the presence of a well-recorded or well-sourced ambience, but its absence may well be noticed. The majority of the work that goes into a production is invisible to its audience — the care and attention placed on clear audio is rarely celebrated, despite being an essential part of the experience.

Voice-Overs and Commentary

Aside from recording sound in the field, a documentary may require you to record an audio commentary. The narrator may, at times, appear on screen, or they may be completely disembodied; they may deliver their material deadpan or with personality, interacting with the visual element of the film. Either way, audio commentary needs to be clear and crisp. As with all dialogue, an audience will not tolerate inaudible or muffled narration. Even if the recording is not perfect, it must be clear.

To accomplish this, a high-quality desktop microphone should be used but, if you do not have access to such equipment, you will have to utilise the resources you have at hand. Employing a lavaliere microphone will not give you the same rich depth that a larger desktop microphone will, but the resultant recording will at least be clear. The prevalence of digital content creation, such as podcasting, has ensured that a wide variety of affordable, quality products are available at a range of price points. If you are able fund the purchase of a desktop microphone, this might well prove to be key investment.

To help in capturing quality audio you should:

- Speak clearly and slowly into the microphone. An accent is fine, but your audience must be able to understand you.
- Be prepared to dislike the sound of your own voice. Everyone hears their voice differently to how the outside world hears it. Whilst it is unusual to hear your recorded voice, you will quickly acclimate to how it sounds.

- Record in a room that does not echo — empty rooms, or rooms without a lot of furniture, will add echo to your voice that you cannot remove in post-production. Conversely, echo, if it is desired, can be added during post-production.
- If you cannot find a space that does not produce echo, create one. Sitting under a table, with duvets draped around it, will create a small space in which your voice will not echo.
- If at all possible, record each line in your script several times and get to know the idiosyncrasies of your voice. Do you raise it at the beginning of a sentence when reading from a script? If so, listen out for that and re-read your line. Did your voice crack? If so, re-read your line.
- Record your voice-over in one sitting (but not one take). Recording over several days will mean that atmospheric changes and subtle (but audible) variations in your delivery tone or pitch will create an uneven commentary track, which may distract your audience.
- Use a pop-filter — these inexpensive pieces of equipment will filter out pronounced 'p' sounds, which your microphone may pick up.

15. Light

Light is important. To some, it may even be the single most important aspect of the filmmaking process, something to be laboured over in the name of aesthetic beauty or intellectual symbolism. For others, it is a variable that requires only as much input or direction as is required to produce a piece of functional, usable footage. In raw moments (those that require no staging), lighting and composition often do not matter. The footage captured on 9/11 is not made any less effective by its lack of controlled lighting. Real moments, captured fleetingly, which cannot be repeated, have an inherent magic which transcends aesthetic beauty. But when a documentary-maker pre-plans a specific scene, be it a sit-down interview or a re-enactment, an audience may expect a more thorough and considered approach to the visual language (and use of light in particular) that is employed. Shot composition can play a large part in this, but so too can the effective use of lighting.

This chapter will provide a foundation designed to facilitate your own experimentations with light. It will provide you with the core knowledge you need to begin understanding light on your own terms, as well as the key knowledge you need to light your shots pragmatically, and the building blocks to begin experimenting with it in more imaginative ways.

Core Rules

Your camera is a light-sensitive device. The more light that enters your camera, the less your device will need to compensate by opening its aperture or increasing its ISO setting. Whilst adjusting the f-stop on a camera or increasing the ISO setting can produce desirable results, they can also alter the image you are capturing in undesirable ways. For example, opening the aperture (reducing the value of the f-stop) will allow more light to enter the camera, but it will also create an image with increasingly shallow focus. This may be the desired effect in some instances, but certainly not all.

The ISO will increase your camera's sensitivity to the light already entering it, but it will also add noise (visual artefacts) to your footage. Depending upon the low-light capability of your camera, this can reduce your image quality a marginal amount — or a very significant amount. Older, entry-level DSLRs and older or inexpensive modern smartphones, for example, produce very noisy, poor-quality images when the ISO setting is pushed too high. As a rule, endeavour to keep your ISO as low as possible, only pushing it higher when conditions necessitate it.

Whilst circumstances will not always allow it, additional light sources can be used to add light to the principal subject within your frame. If you are interviewing a subject, additional light can be used to bring out the details in their face. A well-lit subject will draw your audience's attention to it. This can be accomplished by ensuring that your subject is always facing your main light source. LED light panels are ideal for this task.

Remember: increasing the size of your aperture (decreasing your f-stop value) will let more light into your camera, but create a shallower depth of focus. Increasing the ISO on your camera will make it more sensitive to light, but the higher you set the ISO, the more noise will be introduced to your footage. Depending upon your camera, there will come a point when footage quality degrades noticeably or becomes unusable. Use additional light sources to highlight your subject. Ensure that your subject is angled towards your main light source.

Hard Light and Soft Light

There are two different types of light that are available to you. Hard light (which comes from a single, bright source) creates hard, angular shadows; soft light (which is emitted from a diffused source) creates soft, gentle shadows which wrap themselves around surfaces.

Hard light is a form of bright, unfiltered light. A hard light source, such as the midday sun or a naked filament light bulb, will project a lot of light onto an object, hitting one surface or side, and create angular shadows. This form of light will, on a human face, create areas of darkness which can make the person's features appear harder or more haggard. Shadows may be created around the eyes, for example, or the nose might project a large shadow across much of their face. If you wish to create a sense of menace or imply a negative emotional state, such effects might well be desirable. When working with subjects in the field on a bright, relatively cloudless day, you will need to be prepared to utilise (or compensate for) hard shadows.

Soft light, on the other hand, tends to come from a diffused source and, as a result, the light is more likely to wrap itself around a subject rather than create a stark array of shadows. Soft lighting can bring out nuance and subtlety in facial features, presenting an image that is less harsh in its appearance. Soft light can be created by taking a hard light source (such as a light bulb) and bouncing the light off another surface before it hits your subject. A light reflector, a relatively inexpensive piece of equipment, can be used to achieve this. Hard light can also be filtered through a diffuser, another inexpensive piece of equipment, which will turn it into a soft light source. Whereas hard light comes from a single, powerful source, soft light comes from many different points at the same time, illuminating an object or subject from different angles simultaneously — as a result, shadows are far less pronounced. On a cloudy day, the sun's light is dissipated across the clouds, transforming hard light into soft light.

For dramatic productions, the importance of lighting can hardly be understated. Learning to paint a scene in colour and shadow is an art form unto itself. You need to understand that light remains important, even if the need to control it is typically much reduced compared to, say, a stage play.[1]

A number of documentaries have greatly benefited from careful lighting. *Confessions of a Superhero* features an admirable mix of fly-on-the-wall reportage combined with carefully lit interviews. In the 'real world' scenes, the lighting is situational. On set, however, it is carefully managed, providing a controlled (and very beautiful) setting in which

---

[1] David Landau, *Lighting for Cinematography* (New York: Bloomsbury, 2014); Blain Brown, *Cinematography: Theory and Practice — Image Making for Cinematographers and Directors* (New York: Routledge, 2016); Mercado, *The Filmmaker's Eye*; and Sijll, *Cinematic Storytelling*.

the films' subjects can reflect upon their lives. The controlled lighting relates to some aspect of the subjects' inner thoughts or their life journey. Jennifer 'Wonder Woman' Wegner, for instance, is cast in soft light which gently wraps around her; Maxwell 'Batman' Allen, on the other hand, has hard light (and deep, angular shadows) projected onto him. The difference in the way this pair is lit speaks to the themes each represents within the film. Wegner is depicted as forthright, honest, and kind, and the lighting in her interviews reflects that. Allen, however, is depicted as a much more complicated character, ferocious when angered and liberal with the truth; an enigma who is one part kind and relatable, one part dangerous and deluded. The use of lighting for both subjects is thus coded with meaning. Gentle and abrasive, soft and hard; light and subject are unified.[2]

Even in real-world settings, it would not be unusual for a filmmaker to supplement the light that they find. An LED light attached to the top of your camera can provide enough light to illuminate a subject's face when shooting in the field. Typically, you will position your interview subjects so they stand in front of an interesting background; rarely, however, will the available light complement your choice precisely. A simple LED light panel will allow you to illuminate the subject's face, wherever they are positioned, allowing you to choose a backdrop without being limited by the pre-existing lighting you find in a space. Light is important, and you will need to ensure there is enough to illuminate your subject; you do not need to become a world-class cinematographer, but you do need to understand that there is a relationship between your subject and the light around them. A basic (but important) rule is ensuring that your subject's face is always lit, either by a natural light source or an artificial one.

Make sure your subject is facing towards your main light source. If the main light source in a scene is behind your subject, they will be backlit. In such a setup it can be difficult to bring out details on the subject's face and, depending on the strength of the backlight, either the background or the subject's face will be heavily over- or underexposed. In Figure 47, a subject is photographed in front of the setting sun. The camera is set to expose correctly for the sky. The result is a subject who is rendered almost entirely as a silhouette.

---

2   *Confessions of a Superhero*. Directed by Matthew Ogens. Toronto: Cinema Vault, 2007.

Fig. 47. Backlit by the setting sun, the sky is perfectly clear and detailed whilst the subject is cast into shadow. To bring out the subject's features, a separate light source, aimed at them, would have been required.

Had the camera's settings been altered, to expose correctly for the face of the subject, the background of this image would have been entirely white. The solution to this scenario is the introduction of another light source, this one placed in front of the subject (lighting their front and their face). By applying a light source to the subject, the detail and texture of their appearance would have also been captured alongside the detail and texture of the sky behind them. Alternatively, the photographer could have altered the subject's position, rotating them so that the diffused light from the cloud-filtered sun lit their face. Doing this might have negated the need for a second light source altogether. However, the dramatic view of the sky would have been lost due to the subject and the photographer changing their position.

Perhaps the single most aesthetically useful time for a filmmaker is 'magic hour', the hour before the sun sets. At this time, the sky produces both hard and soft light — particularly the latter as the sun dips towards the horizon. This can create a beautiful effect in which scenes are well lit, but are not dominated by the type of stark shadows that might be produced by the naked sun at other times of the day. 'Magic hour' is a relatively short window of time, however, and though the results of shooting at this time can be striking, it may not be practical to shoot only during this limited window.[3]

---

3   Fenton, *Cinematography*.

Wherever you film, it is your responsibility to understand the lighting conditions that are associated with that space. If possible, you should visit an area at different times of the day, making notes about the types of light and shadow that are present. Note moments when it would be particularly advantageous to shoot for a particular effect. You should also note the limitations of a space's natural light and anticipate any additional lighting needs that may occur as a result. When it is not possible to acquaint yourself with a space ahead of time, ensure that you arrive on location with some way to light a scene or your subject appropriately. This can be simple (an LED light mounted on your camera) or more complex, with lights fixed on their own tripods that can be positioned independently of your camera. The former solution will allow you to create usable footage; the latter solution will allow you to create visually dynamic footage.

Lights and Lighting

Lighting setups come in many shapes and sizes, ranging from the elaborate and powerful to the small and simple. Your lighting needs will very much depend upon what you wish to achieve with your project. A small LED panel should be considered a near-essential purchase. These lights can be easily attached to the top of most DSLRs and, for a basic model, are inexpensive, starting as low as $20 and becoming more expensive as they increase in luminosity and other features.

By adding a light panel to the top of your camera, you will create new opportunities to shoot subjects in low-light conditions. Whilst the light provided by such panels is unlikely to help you to create a cinematographic masterpiece, it will allow you to film in otherwise problematic conditions. Over time, lights can be acquired piecemeal and added to your kit. A small light mounted on your camera is an essential first step, but LED light panels mounted onto stands will provide you with significant flexibility when interviewing a subject. Panels with high-quality rechargeable batteries add to the cost of such lights, but increase their practical usage significantly.

Fig. 48. This LED panel cost less than $60 and can be mounted to a stand. It comes with a number of different filters, which can be used to defuse the light whilst increasing or decreasing the light's colour temperature.

In the field, natural lighting should always be the filmmaker's first point of reference — what can be accomplished with the natural light available at a given time on a given location? There are occasions, however, when a more considered approach to lighting in the field must be taken. Re-enactments or complex set pieces, particularly where any noteworthy level of expense is incurred through their staging, will likely require a degree of forethought with regards to how they are lit. Even if the intention is to use natural lighting as much as possible, unexpected weather conditions may render this more difficult than anticipated. In such instances, portable field-lighting solutions are available. These typically involve LED light panels that can be attached to stands, allowing them to be positioned independently of one's camera. Such lighting setups are more expensive than small camera-mounted LED panels, but they provide significant freedom should you wish to stage more complicated, cinematic sequences in the field.

More controlled environments, particularly those to which a filmmaker has regular access, can create opportunities to employ more permanent lighting setups. Whilst the field lighting above can be used to light a studio-style space, a set of soft-box lights, which generally require a lot of power and are thus more suited to indoor environments with access to a mains electricity supply, can create effective soft lighting. Unlike the filament lights, which can flicker noticeably when filmed, these lighting solutions provide continuous light which is filtered through a diffuser. These lights are more cumbersome than their LED counterparts and their reliance upon mains electricity limits their versatility. For indoor projects and studio spaces, however, they can be particularly useful.

The prices for such setups vary widely, with basic LED panels available for less than $20 and more advanced LED systems available for more than $1000. As with all of the tools discussed in this volume, it is not always necessary to spend very large sums of money to buy the best equipment. Rather, you should focus upon using whatever equipment you possess effectively. An expensive lighting rig will not necessarily result in a well-lit scene. Likewise, inexpensive lighting solutions do not necessitate poor results. The careful and considered use of one's resources, whatever they may be, is the critical factor. Natural light is perhaps the most valuable resource available.

Lighting Quick-Reference Guide

To ensure you subject is sufficiently lit, angle them towards your main light source.

- Hard light comes from a single source (such as the sun or an unfiltered bulb) and creates hard, angular shadows.
- Soft light is emitted from a broader area (such as the sun shining through clouds) and creates softer shadows and contours.
- If you wish to backlight a subject, or place them in silhouette, place them in front of your main light source and adjust the exposure settings on your camera until you capture the desired effect.
- A basic LED light panel can be fitted to most DSLRs and will allow you to create usable footage in a wide variety of situations.
- More complex lighting setups involve lights that can be placed independently of your camera. LED light banks can be powered by batteries, allowing for versatile lighting kits that

can be taken into the field with comparative ease. A soft-box solution can be employed in permanent or semi-permanent indoor spaces with access to mains power.

16. Camera Movement

Moving the camera (and getting usable footage) is a deceptively difficult task. Handheld DSLRs and smartphones, thanks to their small size and lightweight nature, absorb the natural vibrations of the user's hands, arms, and chest. This can result in footage that is distractingly unstable. The natural vibrations in your hands, your arms, and your fingers can easily transfer into your device, creating off-putting footage which vibrates or shudders in unnatural ways. Holding a camera directly with your hands should thus be avoided.

There are, however, a number of solutions available if you wish to move your lightweight camera. These solutions assume that you do not have the budget to purchase sophisticated stabilisation kits and will instead aim to provide work-around solutions using the types of equipment you are likely to own as a part of your basic kit. These solutions use this basic equipment in imaginative ways to achieve effects that normally require specialised equipment.

Going Handheld

One quick and reasonably effective way to compensate for camera shake is to add more weight to your camera. This simple addition will help to compensate for the natural vibrations and movements that your hands introduce to your equipment. A tripod (with its legs closed) can be used as a rudimentary type of stabilisation rig — rather than holding your camera directly, instead grip the folded tripod to which it is attached. The additional weight will help to reduce the amount of shake that you introduce to your footage, whilst the tripod itself will absorb some of the vibrations and movement that can make handheld footage so unstable.

Understanding how your body works in relation to your camera can also help you to add a layer of stability to your footage. Every time you inhale, your chest rises and the position of your shoulders, and therefore your arms, changes. When using a camera handheld, you should thus be aware of your breathing and endeavour to control it, limiting the movement of your chest and arms. Shorter, more controlled breaths can help significantly and, combined with the additional weight and stability added by a tripod, will help you to capture more usable footage.

When particularly stable shots are required from a handheld camera, it may be necessary to hold your breath in order to ensure minimal movement in your chest and arms. If your arms are outstretched, they will be in a position of tension — inevitably, they will get tired and that will, sooner or later, result in them moving or vibrating in a way that will make your footage increasingly unsuitable. To compensate for this, bend your arms at the elbow and tuck them into your ribcage. This will ensure that the weight of the rig will be passed into your body with less strain on your arms, allowing you to hold your camera in a steadier position for longer. Combine with holding your breath (or controlled breathing) for the best results.[1]

More specialised equipment — rigs — can greatly increase the ease with which you can move your camera. A gimbal adds moving parts and counterweights to your camera's support mechanism, allowing some degree of camera shake and wobble to be absorbed by the device. These devices are particularly useful for moving the camera, allowing an operator, with practice, to track a subject and collect usable footage.

A C-grip allows you to hold the camera from above, turning it in a number of different directions, without ever having to touch it directly (see Figure 49). The distance of the camera from the handle, coupled with the shape of a C-grip, helps to remove the shake that would otherwise be introduced by your hands. C-grips are particularly useful when you wish to be able to move the camera freely whilst standing in a stationary position. For example, if you wished to film a skateboarder performing tricks on a halfpipe, a C-grip would allow you to stand close to the action whist moving the camera freely to track the skater. Whilst some version of this type of camera movement could be replicated using a tripod, the camera would only be able to track the skater from a fixed pivot point (the location of the tripod head). In addition, tripods have

---

[1] Fenton, *Cinematography*.

a large footprint, which can make them impractical, even dangerous, to use in close proximity to a fast-moving subject.

Fig. 49. A homemade rig, assembled over time from inexpensive but effective component parts. A C-grip forms the basis of it. Cold-shoe extenders allow for external accessories, including lights and microphones, to be added to the rig. This is a handheld setup that has been attached to a tripod for stationary shots without needing to be disassembled.

It is also possible to acquire, even build, an inexpensive rig that combines different stabilisation elements which allow you to operate quickly and efficiently in fast-changing situations. Figure 49 shows a rig using a C-grip as its basis. It easily connects to a tripod whilst a range of cold-shoe extenders (simple metal devices that allow accessories to be attached to the socket where a camera flash would normally be attached) allow for the addition of external microphones, lights, and so on. By folding up the tripod and placing it across one of the operator's shoulders, this rig transforms into a shoulder mount. Such setups are less effective than dedicated stabilisation rigs, but they can be constructed from inexpensive materials over a period of time. For budget-minded filmmakers, such solutions are effective and versatile.

Handheld Tracking

If you want track a subject with your camera through a physical space then you will need to practice how you walk. Most people have a natural amount of bounce in their step — the human eye and brain compensate for this so we are unaware of it as we walk. The camera, however, will capture this bounce in uncomfortable, sudden shifts along the vertical axis.

Fig. 50. Tracking shot captured in New York by a camera operator following two subjects. *Looking for Charlie* (0:30:58–0:32:37).

In order to track a subject through space with a handheld camera, you will need to modify the way you walk. The final part of the step — literally, the spring in your step — needs to be excised. As you walk, notice that the heel of your foot lifts up before your toes spring your foot and leg into the air. When tracking a subject with a handheld camera aim to raise and lower the heel and your toes evenly. Bend your knees as you walk to ensure that you do not bob up and down as you move. This will result in a strange-feeling, flat-footed walk — but it will help create much smoother footage.

To gain additional stability for a complicated camera manoeuvre, fold the tripod so that its legs sit perpendicular to the camera, forming a horizontal bar that extrudes from the back of your camera. Place this bar (your folded tripod legs) onto one of your shoulders. With your hands, hold the end of your tripod closest to your camera. You now have a makeshift shoulder rig which, coupled with a bounce-less walk, will allow you to track people through physical spaces in a comparatively stable manner. Significant practice will be required to perfect your 'tracking walk'.

Fig. 51. A folded tripod placed across the shoulder can serve as a crude shoulder stabiliser. When using such a setup, walk with bent knees, raising and lowering your feet so that they remain parallel to the ground. Do not push up using the ball of your foot to avoid ruining your shot with a bounce.

Camera Pans and Tilts

A common but effective shot that you may want to employ is the camera pan: the camera remains stationary, but looks (pans) around a scene along the horizontal axis. This can be a particularly effective way to take in a scene that is too large to be effectively captured in a single, stationary shot.[2] It is a relatively easy effect to create as it requires you to loosen your tripod head just enough so that your camera is able to look around freely when you pull on the control handle. Like almost all types of human-controlled movement, however, unwanted shake and vibration can be introduced.

To compensate for this, loop an elastic band around the control handle on your tripod and pull on this (rather than on the handle

---

[2] Elliot Grove, *Raindance Producers' Lab: Low-to-No Budget Filmmaking. Second Edition* (Burlington: Focal Press, 2014), pp. 53–60.

directly) to create camera motion. The elastic band will absorb shake from your hand and, assuming you pull it at a steady rate, it should provide you with a smooth pan. Practice, however, is essential. As you drag the camera around, you may well find that, as your tripod head loosens, the speed of your pan increases. In order to compensate for this, you will need to practice the motion, gaining a sense for when the movement of your tripod head starts to speed up (or slow down) and compensate for it appropriately.

Camera tilts can be accomplished in practically the same manner. If a camera pan describes the motion of a camera as it looks from left to right (or right to left), a tilt describes a camera as it swings along the vertical axis. To accomplish this move, loosen the tripod head. Again, loop an elastic band around the control-handle, this time pulling it so that the camera tilts in the desired direction. Once again, practice the motion, learning when your tripod head will loosen or tighten to an undue degree.

Dolly Shot

A dolly shot is achieved when a camera is placed on a moving object — this, in theory, should provide you with a very smooth shot as the camera tracks closer to your subject. Dolly shots are, however, deceptively difficult to achieve. In professional productions, dollies are often placed on tracks and pushed by several members of the crew. This is a time-consuming and expensive way of creating such a shot.

You can reduce the expense — but not the time — by placing a camera on an office chair or similar device. When shooting *Looking for Charlie*, a camera was placed on top of a suitcase and then slowly wheeled towards its target to create a tracking shot. This solution worked, but it was time-consuming. An entire unit had to dedicate themselves to the task of capturing a single, simple tracking shot which, in the end, took upwards of an hour to shoot and resulted in only a few seconds of screen time. Such budget-minded solutions also carry risks. A camera placed on top of a suitcase is liable to fall and break. If the surface over which a makeshift dolly is moving is uneven, a significant amount of distracting shudder might be introduced to a shot, ruining the take.

Makeshift dollies also have a tendency to wander off course — without tracks they can be difficult to move in a perfectly straight line.

These problems are not insurmountable, but they do require time, patience, and practice. Set aside a sizeable amount of time to achieve a dolly shot. You will in all likelihood need to practice the shot and, if you are working with others to achieve it, you will all need to work in an effective, collaborative manner. All of this requires significant patience, not only on the part of the director but all of those working to accomplish the shot. Many takes are likely to be required and repeated failures can lead to frustration.

Considering the difficulty of attaining tracking shots, inexperienced filmmakers should consider the effort/reward ratio involved in a given shot. If the dolly shot communicates something to the audience that would not be easily replicated with another type of shot then, by all means, work towards achieving it. But do so understanding that it will likely take you longer (and require greater patience) than you imagined. The results, however, can be really quite effective when a successful take is finally captured.[3]

As difficult as a dolly shot can be, there are some hacks you can employ:

Tripod Dolly: not only can your tripod act as a rudimentary camera rig, it can be used to create a type of faux dolly effect. This can be accomplished by loosening your tripod head so that your camera is free to move on its vertical axis (up and down). By stepping forward so that your tripod pivots on to its front two legs, you will be able to move the camera forward in a comparatively smooth manner (see Figures 52 and 53).

Makeshift Dolly: what is a dolly? Potentially anything with wheels, on which you can place your camera. Whether or not it is effective depends on a number of factors. Is your dolly going to be moving over a smooth enough surface; is it stable enough; have you the time and patience to repeat the shot, over and over, until you think you have captured precisely the effect that you want?

---

[3] Barry Andersson, *Filmmaker's Handbook: Real-World Production Techniques. Second Edition* (Indianapolis: John Wiley and Sons, 2015), pp. 50–52.

Drone Dolly: this is an emerging solution to the dolly shot; high-quality video drones are now able to capture smooth tracking footage which, if used in tandem with a skilled pilot, can open up many possibilities for creating dynamic, moving shots. The main issue with drone technology is that, at this stage, it remains expensive, with even modest video drones capable of capturing usable footage starting at approximately $600, with more sophisticated devices costing upwards of $1,500. Whilst there are a range of inexpensive drones which claim to be able to capture high-definition footage for less, these should typically be avoided. Cheap drones tend to have poor-quality cameras, which are mounted in a way that fails to compensate for the vibration created by the vehicle's motors. As drone technology continues to improve, look for more effective and affordable solutions appearing on the market.[4]

Train Dolly: a simple, low-cost, but effective solution to create an environmental dolly shot is to place your camera flush against the window of a moving train, subway, or tram car. If the vehicle is moving through an interesting urban environment, it is possible to create dynamic, moving shots which can greatly add to your production. Rush hour and other busy periods should be avoided, and shots tend to be most effective when the vehicle is moving at a slow but steady pace through a spatially interesting area. If you are able to coordinate all of these factors, however, this is an inexpensive and easily actioned method of capturing environmental dolly shots.

---

4   Eric Cheng, *Aerial Photography and Videography Using Drones* (Berkeley: Peachpit Press, 2006).

Tripod Dolly

Figs. 52–53 The tripod dolly: the tripod's front legs remain stationary as the entire set up is pushed forward. The tripod's head is loosened so that the camera can remain perpendicular to the ground.

17. The Two-Page Film School

Fig. 54 Watch the video lesson on conducting interviews. https://hdl.handle.net/20.500.12434/c9b0163c

If you are setting out to make your first film, the amount of practical advice available can feel overwhelming. In this book, much of this advice has been has been distilled down to the basics, but it can be distilled yet further. As *Sin City* (2005) director Robert Rodriguez once put it, 'everything you need to know about filmmaking… You [can] learn it in ten minutes.' That is a generous assessment, but Rodriguez was really referring to the technical aspects of the production process, something he was keen to demystify throughout much of his career. Rodriguez believes it is possible to learn the necessary filmmaking techniques in just ten minutes because he understands that there are a core number of rules which, if followed, will allow for the capture of competent, usable footage. Everything else is practice, dedication, and imagination.

Rodriguez tells us that mastering the technical aspects of the process is not the time-consuming part. It is developing one's own voice and vision that takes time; indeed, Rodriguez spent most of his childhood learning how to be a filmmaker.[1] He was not, however, willing to allow the intimidating mechanics of filmmaking stop him from transitioning from hobbyist to professional. In that spirit, this chapter distils the core lessons of the preceding chapters into a simple, two-page film school — the ultimate distillation of the preceding chapters' practical advice. In the above video lesson, we will take you step-by-step through the process of setting up a one-camera interview. Below, we have curated the core lessons you need to remember when you are in the field:

1. Set your camera up to shoot at 24 fps and, if you can change the shutter speed, set it to 1/50 or 1/48.

2. Leave the white balance on automatic unless you want to change the colour profile of the image you are capturing.

3. The more you zoom in to an object (using an optical zoom), the more you will flatten your footage — objects in the distance will appear much closer to those in the foreground the more you zoom in.

4. Download a light-meter app for your smartphone. This will allow you to aim your phone at a scene and it will then tell you the settings that you need to put into your camera. If you are using a DSLR as your main camera, make sure you lock the shutter speed (in both the light-metre app and the camera) at 1/50. The app will then tell you what settings you need to change on your camera in order to capture correctly exposed footage.

5. Have your subject's face angled towards your main light source.

6. To backlight a subject, place them in front of a bright light source and adjust the exposure on your camera until you achieve your desired effect.

---

[1] Robert Rodriguez, *Rebel Without a Crew: Or How a 23-Year-Old Filmmaker with $7000 Became a Hollywood Player* (London: Penguin, 1996).

7. Never rely on the internal microphone in your camera. Get a good-quality lavaliere microphone (which start as low as $30–50) which can record directly to your smartphone. For clearer run-and-gun sound (when you cannot mic up a subject), buy a directional microphone that you can attach to your camera.

8. Always stabilise your footage. Use a tripod for a stationary image or some kind of rig (including a folded-up tripod) to allow you to go handheld.

9. Double or triple check to make sure you have focused on the correct part of each frame you are shooting.

10. Compose your shots using the 'rule of thirds' as your guide.

11. Watch DVDs with director commentaries — every one of them is a micro film school.

12. Use your limitations to your advantage. Problems require imaginative solutions to overcome them. Respond with the equipment and resources at hand in the best way that you can manage. In other words, think on your feet and be prepared to adapt. You do not need expensive equipment to make a compelling film. You need to use your resources, whatever they are, in the most effective and imaginative way possible.

18. Post-Mortem
Collaborating with Students to Make a Documentary about the Election of Donald Trump

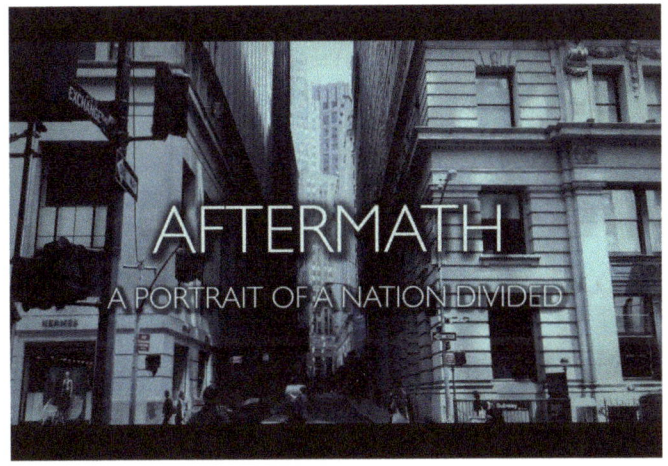

Project: If He Wins (Working Title)/Aftermath: A Portrait of a Nation Divided (Final Title)

Anticipated Running Time: Approximately 5-8 Minutes.

"Rationale: The 2016 presidential election is proving to be a particularly divisive affair, with the success of Donald Trump suggesting a change in the political dynamics in the United States. The result is a historic electoral process in which the candidates (and their personalities) are threatening to overshadow the electorate. As a result, this film will aim to capture a snapshot of how ordinary citizens in New York, an important city to both candidates, are responding to the changing political landscape. This film will present the views of its respondents in an honest and transparent way, whatever they are."

In May 2016, we travelled with ten of our students to New York in order to create a short documentary about the unfolding presidential election. As outsiders, we wanted to capture a snapshot of the city's mood, a portrait of how people were feeling about the divisive election and, in particular, Donald Trump's spectacular rise to prominence. Our core concept was simple: ask the residents of the city what would happen if he won. We wanted to create a short film that reflected the mood we discovered. We saw ourselves as observers, not provocateurs.

This election seemed to demand particular attention. Having filmed in New York previously it made sense to revisit that location, although, as a democratic stronghold, it was a potentially problematic choice. Still, we anticipated being able to capture a multiplicity of perspectives. Ideally, we would have travelled to several locations, in different parts of the country, and spoken to a wide cross-section of people. Our resources, however, put a strict limit on our ambition. We would make New York our case study and attempt to correct for its Democratic bias. New York may have been a blue state but, we reasoned, supporters of Trump would nonetheless be present.

From the outset there were three major factors that would help to shape our thinking throughout the filmmaking process. Firstly, the film would be released on platforms such as YouTube; it would likely be consumed as part of our audience's regular diet of bite-sized content. Secondly, we did not want to appear in the finished film; this should be a story by and about the people of New York. Thirdly, we wanted to reflect the uncertainty of the moment by having our subjects speculate about what the future under a (then) theoretical Trump presidency might look like; uncertainty mirrored by speculation about the unknowability of the future.

The desire to release the final piece via online video streaming services meant that we had to pay attention to the ways in which media was consumed on such platforms. To that end, we aimed to create a film that would fit easily into YouTube viewing patterns. It had to be long enough to interest people, but not so long that it would impose upon someone's day — a five-to-eight-minute burst of concentrated discourse. As we did not deem it appropriate to appear in the film, to include a commentary track would, we felt, likewise pull attention away from our subjects, as well as adding undue length and complexity to a project that did not require either. Problematically, however, remaining off-camera would also serve to obscure our biases from the audience. By choosing to remain off-camera, we knew our film might present the illusion of greater objectivity. The filmmaker always crafts the truth that appears in their work and, whatever problems are introduced when they choose to appear on screen, their presence at least reminds the audience that they are watching a subjective piece loaded with authorial bias. Still, it was important to us that we make a film that would be built exclusively around the views, ideas, and perspectives of the people of New York. Reality was the real director of this project and so it was real life, rather than ourselves, that needed to appear on screen.

Of course, reality has to be framed. Asking our interviewees to simply give us their impressions of the election would be unlikely to lead to a particularly coherent, or deep, set of discussions. As a result, we constructed an interview questionnaire which was designed to encourage our subjects to reflect upon the nature of the country, and where it might be going in the future.[1] The 2016 election was nothing if not an event filled with speculation about the type of country the United States was, and the type of country it wanted to be. To capitalise upon that existential dimension, our questionnaire culminated with a simple question: 'What happens if Trump wins?' This question became our central organising principle during the early planning stages of the film and, consequently, *If He Wins* became the project's working title.

Once our core concept and questionnaire were written, we set about the task of planning our shoot. As we could not predict how our interview subjects would respond to our questions, or even who

---

1   For discussions on the process of designing oral history projects, see Paul Thompson, *The Voice of the Past: Oral History. Third Edition* (Oxford and New York: Oxford University Press, 2000), pp. 222–308; Ivan Jaksic, 'Oral History in the Americas', *The Journal of American History* 92 (1992), 590–600; Alistair Thomson, 'Four Paradigm Transformations in Oral History', *The Oral History Review* 34 (2007), 49–70.

they would be, it was difficult to imagine what our final film would look like. We could, however, plan how we would go about gathering a range of different perspectives by identifying locations within the city where we might expect to encounter different demographics. Brooklyn, Wall Street, Coney Island, and Harlem were selected and a production schedule was built around visiting those locations.

Pre-Production

To ensure an orderly production we carefully planned our week-long schedule, accounting for where we would shoot, when we would be on location, how long travel between locations would take, and so on. Learning from our last trip to New York, we were careful not to overstuff our schedule. Aside from planning the shoot, pre-production was also the period during which we reviewed and assessed the equipment available to us:

1. A Nikon D5500 and three lenses: 18–55mm, 50mm, and 55–200mm. The 18–50mm lens had proven to be a capable workhorse in the past and would prove, once again, to be ideal for capturing a wide range of environmental footage. Its variable aperture size would help to provide a broad depth of field, which would keep moving subjects in focus. The fixed 50mm lens was an ideal lens for shooting interviews, with a maximum f-stop of 1.8 creating shallow-focus shots which fixed the viewer's attention on the interviewee. The 55–200mm lens would allow us to compress spaces in our shots, or capture moments that would otherwise be out of range for our other lenses.

2. A Nikon D3100 with an 18–55mm lens. Broadly comparable to the D5500 in daylight conditions, the D3100 is an early-model DSLR which struggled in low-light. Being very familiar with this device, we understood its limitations and quirks, allowing us to circumvent its limitations in order to put it to the best possible use. Despite it being significantly inferior to the D5500, it provided the crew with a solid second camera, particularly in situations where high-quality natural light was available.

3. Acquiring a third camera proved to be more problematic. Beyond funds for our trip to New York, *If He Wins* did not have a budget upon which we could draw to purchase (or even rent) additional equipment. Our solution was to use an iPad, recognising and compensating for its limitations as much as possible. Whilst dedicated camera equipment is almost always the preferred option, the video-capturing ability of devices such as the iPad has improved significantly in the past few years. Smartphones and tablets are nowhere near as versatile as a high-quality DSLR, but that does not mean that they are not capable of capturing high-quality footage in the correct circumstances. Our online streaming model, which anticipated people viewing the film on smartphones and similar small-screened devices, further justified the use of such equipment.

4. Tripods were sourced for each camera. For the iPad this required a tablet-to-tripod mount. A guerrilla tripod, a small device with posable legs that allows camera equipment to be mounted in a variety of unusual locations, was also sourced for the project. To record audio, two lavaliere microphones were acquired. A microphone that could be mounted to our lead camera (costing approximately $80) was also included in our manifest. The lavaliere microphones were connected to smartphones to record interview audio.

5. Release forms, to allow us to use the footage that we captured, were created, along with multiple hard copies of our production schedule.

6. A 360° camera. A colleague at our institution had recently held a session designed to inspire the creation of 360° and virtual reality films. Intrigued by the concept, we borrowed a 360° camera in order to experiment with it on our shoot. Our inexperience with the camera meant that we had no expectation that we would be able to capture anything worthwhile using this equipment. Whilst we believe we were correct not to shower undue (and unearned) attention on this new device (making a 360° film was, at best, a secondary concern for us) the decision to use the camera provided us

with an opportunity to successfully assemble our first virtual-reality film following our return.

To maximise the quality of the footage we would capture on the iPad, we utilised an app called FiLMIC PRO, which allowed for the device to record video at a range of frame rates, including the cinematically desirable 24fps. The app also allowed us to adjust exposure and focus separately, a pair of functions that are normally combined in the device's standard camera app. Despite the additional functionality we were able to eke out of the device, its dynamic (colour) range could not match that produced by our DSLRs and, as a result, particular attention had to be paid to the iPad footage during the post-production process. Still, the iPad proved to be a competent third camera. The footage captured by it is difficult, if not impossible, to identify in the final production; as a result, we were able to divide our crew into two separate units, each able to carry out different tasks simultaneously. Whilst cameras one and two (the Nikon D5500 and D3100) would be used primarily to shoot interviews, a second unit could use the iPad to capture environmental footage, allowing us to maximise our time at each of our chosen locations.

Production

Day One: Our first day of production was spent familiarising the crew with their roles. To that end, we spent the first day shooting in Central Park, engaging in a pop-up seminar where we talked through our own feelings about the election and took part in other team-building activities. Several games of Frisbee, some work on a promotional video for our institution, familiarising ourselves with the equipment; none of this led to the creation of any substantive footage, but it did help our crew come to grips with the larger task at hand and to settle into the process.

Day Two: Following our first day in Central Park, we travelled to Brooklyn where we scouted a suitable location to capture our first set of interviews. Setting up our equipment, we approached passers-by, telling them about our project, and inviting them to participate. Convincing people to appear on camera was not easy, however. Many potential subjects seemed interested in our project but were, understandably, reluctant to speak to a group of strangers (on camera, no less) about their political beliefs. Despite having found a suitable location with reasonable foot traffic, it was not always easy interrupting peoples' days. Many were simply not willing to engage with us. This, we completely respected. Many invitations were offered and turned down but, over the course of the day, we were able gradually to acquire a bank of interviews. This included one brief on-camera discussion with a Trump supporter — the only one we were ultimately able to capture on film.

Day Three: Our second shoot took place at Coney Island, a quirky, eccentric, and anachronistic beachfront arcade. Again, we encountered some difficulty in acquiring interviews but a more noteworthy pattern was starting to emerge in the material that we were able to collect. Though we encountered Trump supporters who were interested in talking to us about their political beliefs, they had little interest in appearing on camera. One individual in particular spent a considerable amount of time watching us shoot, engaging us in discussions about the reasons he would vote for Trump, but he was unwilling to speak on camera. Despite capturing a number of quality interviews with Trump critics at Coney Island, we had failed to capture a single Trump supporter on film.

Day Four: Rest Day.

Day Five: By the time we began shooting at our third location, Wall Street, the growing imbalance in our material was becoming evident. Wall Street was, we assumed, one of the locations where we were most likely to find Trump supporters. As it turned out, it was extremely difficult to convince *anyone*, pro or anti Trump, to appear on camera at this location. In one notable exchange, a crew member asked a passer-by if they supported Trump. 'Yes,' they answered. 'Would you say that on camera?' the crew member followed up as the passer-by brushed past them. 'Nope,' he shouted back at us.

In another instance, we fell into a conversation with a group of workmen who were happy to talk about the election but unwilling to speak on camera. Of the three, two were openly critical of Trump. The third, however, after a good degree of preamble, expressed support for some of Trump's policies. The discussion was convivial and constructive but they ultimately declined to share their views on camera. By this point it was becoming clear that Trump's New-York-based supporters were reluctant to openly share their sympathies for the candidate or his policies.

By the end of our time on Wall Street we had succeeded in capturing only two interviews. An exhaustive amount of work had gone into acquiring those interviews but they did not reflect the more diverse political views our off-camera conversations had exposed us to. In retrospect, something more should have been done about this; not to force interviews from reluctant subjects, but to somehow represent, on-screen, the reluctance of Trump supporters to speak about their support for him.

Day Six: Our final shoot took place in Harlem and, unlike our recent experience on Wall Street, a wide variety of subjects were willing to share detailed reflections on camera. Whilst our time on Wall Street had been difficult, our time in Harlem was a joy. That is not to say that it was without incident. At one point a young musician approached our group and accused us of treating Harlem like a 'zoo', informing us that we should be spending money, so that we might support local businesses and Harlemites like himself. He then called us all racists and left. It was an instructive moment, which spoke to deeper tensions in the area related to gentrification and identity politics. Later that day, he returned to apologise, explaining that he had been trying to convince us to buy his new CD. We then bought a copy.

With only minimal effort, we were able to attract a range of subjects to our camera in Harlem, each of whom delivered a charismatic and enthusiastic series of responses to our questions. In one instance, we were able to convince the owner of a local business to speak on camera, if we agreed to shoot a short video about their establishment. Despite a pressing schedule, we obliged, happy to pay something back to a community that had been so generous and welcoming. Despite rounding off our shoot with a series of quality interviews, the material we captured in New York reflected only one side of the discourse to which we had been exposed. Balance was an issue that we had become increasingly conscious of, but our principal aim was to allow New York to speak for itself, allowing the material we captured to direct the film that we would ultimately produce. By the time we left New York, however, it was evident that our film would primarily present the views of those who were critical of Trump.

Post-Production

We did not enter post-production immediately. Instead, we chose to wait until the election reached a point when our material could contribute constructively to the emerging discourse. Problematically, Trump seemed, according to our own instincts, to be an unlikely victor throughout much of the election and the footage we captured seemed to reinforce that narrative. As a result, it was unclear what our film would add to the discussion. Following Clinton's post-convention bounce, the chances of Trump winning seemed remote.[2] Provisionally, we decided to return to the material in late September following the first presidential debate.[3]

Events in the 2016 presidential race were prone to sudden and unexpected changes. Following the first debate, Trump's attacks on Alicia Machado, the former Miss Universe winner whose looks he had publicly disparaged, set off a maelstrom of criticism which seemed to signal the start of an unstoppable downward spiral for the candidate.[4]

---

2   Edward Helmore, 'Hillary Clinton Sees Post-Convention Boost over Trump, But Will it Last?', *The Guardian*, 30 July 2016, https://www.theguardian.com/us-news/2016/jul/30/hillary-clinton-donald-trump-post-convention-poll; Alan Rappeport, 'New Poll Reflects a Post-Convention Bounce for Hillary Clinton', *The New York Times*, 1 August 2016, https://www.nytimes.com/2016/08/02/us/politics/clinton-convention-poll.html; Steven Shepard, 'How Big is Hillary Clinton's Convention Bounce', *Politico*, 2 August 2016, https://www.politico.com/story/2016/08/how-big-is-hillary-clintons-convention-bounce-226545

3   The apparent weakness of the Trump campaign was exacerbated further following the debates, which failed to offer any further clarity regarding the place of our film: see Maxwell Tani, 'Hillary Clinton's Debate Surge is Now Clear', *Business Insider*, 4 October 2016, https://www.businessinsider.com/hillary-clintons-polls-debate-winning-2016-10?r=UK&IR=T

4   See Lucia Graves, 'Alicia Machado, Miss Universe Weight-Shamed by Trump, Speaks Out', *The Guardian*, 28 September 2016, https://www.theguardian.com/us-news/2016/sep/27/alicia-machado-miss-universe-weight-shame-trump-speaks-out-clinton; Michael Barbaro and Megan Twohey, 'Shamed and Angry: Alicia Machado, a Miss Universe Mocked by Donald Trump', *New York Times*, 27 September 2016, https://www.nytimes.com/2016/09/28/us/politics/alicia-machado-donald-trump.html; Jannell Ross, 'Alicia Machado, the Woman Trump Called Miss Housekeeping, is Ready to Vote Against Donald Trump', *The Washington Post*, 27 September 2016, https://www.washingtonpost.com/news/the-fix/

Our original question ('what happens if he wins?') could not have felt less relevant.

That was ultimately a good thing. The original framing question was not particularly inspired, and our footage showed that, underneath many carefully considered answers was a deep sense of unease. As a result, we began to rethink how the film would frame the interviews we had collected — as ever, the absence of substantial material from any Trump supporters weighed heavily upon us. The release of the 'grab them by the p---y' tape weighed even more heavily: laughable though it seems now, as we were editing our film we had to consider the possibility that Trump would pull out of the race entirely.[5] Indeed, he might, we reasoned, pull out of the race before we had an opportunity to release our work to the public.[6] So we became reactive.

The original title, *If He Wins*, was thrown out in favour of something more abstract: *Aftermath: A Portrait of a Nation Divided*. Even that title did not feel entirely appropriate. We could not precisely define the aftermath to which we were referring: the aftermath of Trump's divisive language; his candidacy; or maybe his failure to prove himself even vaguely capable of winning? The change in title was a reflection of the confusion of the moment and our own misreading of the political temperature in America. Unexpectedly, it was the silence of Trump's supporters in our piece that ultimately gave it meaning. Like so many pundits and commentators, we had come to labour under the impression that Trump could not win. What we did not realise, and what our film reflected, was the weight of the silent voice in American politics at that moment. This was something that would only become clear in the aftermath of the process.

---

wp/2016/09/27/alicia-machado-the-woman-trump-called-miss-housekeeping-is-ready-to-vote-against-donald-trump/; Peter W. Stevenson, 'The Clinton Campaign Had Been Getting Ready to Drop Alicia Machado on Trump for a Long Time', *The Washington Post*, 27 October 2016, https://www.washingtonpost.com/news/the-fix/wp/2016/10/27/inside-the-clinton-campaigns-anti-trump-surrogate-rollout-plans/

5   For context on the 'grab them by the p---y' tape see "Transcript: Donald Trump's Taped Comments about Women' *The New York Times*, October 8th, 2016, https://www.nytimes.com/2016/10/08/us/donald-trump-tape-transcript.html .

6   Lauren Gamino, 'What Happens if Donald Trump Pulls Out of the U.S. Election?', *The Guardian*, 9 October 2016, https://www.theguardian.com/us-news/2016/oct/08/what-happens-if-donald-trump-quits-presidential-race-election-experts

Aftermath

Fig. 55   Watch *Aftermath: A Portrait of a Nation Divided.* https://youtu.be/bU1wf4UIt-o.

Overall, we are proud of *Aftermath*.[7] We had wanted to create a filmic portrait, allowing the people of New York to create a collective narrative about a specific moment in time. We had wanted to represent the people we met, not manipulate them. Following its release, *Aftermath* generated the type of discussions we hoped to see — we had not set out to be provocateurs, but every documentarian ultimately becomes one. At screenings and online, the film helped to generate discussion, debates and, in some cases, partisan fury. Despite our inability to convince Trump supporters to appear on camera, we acknowledged this at the end of the film and, in that way, gave their silence some degree of weight. The film did not argue that Trump lacked support, only that many of Trump's supporters in places such as New York were reluctant to share their views in an open or transparent way.

We had met Trump supporters but, with only one exception, heard in the film's opening, none agreed to appear on camera — and even that subject said little more about the candidate than is presented in the

---

7   *Aftermath: A Portrait of a Nation Divided*. Digital Stream. Directed by Brett Sanders and Darren R. Reid. Coventry: Red Something Media, 2016.

film. As a result, the lack of balance we had achieved felt appropriate, particularly as Trump's chances of victory appeared to approach zero.[8] We kept the tone of our final comment as neutral as possible: 'Although we met supporters of Donald Trump, they refused to speak to us on camera'. This acknowledgment was an honest reflection of our attempt to attain balance, giving the preceding interviews an additional level of meaning. Beyond the highly motivated and outspoken Trump partisans, *Aftermath* helped to illustrate that support for the candidate was not always boisterously or openly expressed.

To our mind, the silence of Trump's supporters gave them a unique voice in our film. The silence said something, though we did not know what at the time. In retrospect, it echoes loudly. At our first post-election screening, the audience laughed aloud as our final subject, in her charismatic manner, decried Trump and his policies. The expletive thrown in by a passer-by ('F--- Donald Trump!') amplified their laughter. But as our acknowledgement of the silence of Trump's supporters appeared, some members of the audience gasped audibly. There was a sense of palpable shock at the screening. *Aftermath* had not drawn this type of reaction prior to Trump's victory in the election.

Audience members had laughed, but before this they had never recoiled or shown visible signs of shock at this final reveal. After the election, however, that final piece of text seemed to completely reframe everything that preceded it. Before the election, the film had been a comfort to audience members critical of the candidate's policies and rhetoric. After the election, the echo chamber was broken. A new truth (not to be confused with reality) had emerged in the film. Or rather, the weight of interpretation had shifted. The film itself has not changed, but its meaning had. An imbalance that seemed to annoy some audiences prior to the election now appeared to be telling, foreboding even. The hint of an electoral sleeping giant had transformed into a rebuke.[9]

---

8   Chris Cillizza and Aaron Blake, 'Donald Trump's Chances of Winning are Approaching Zero', *The Washington Post*, 24 October 2016, https://www.washingtonpost.com/news/the-fix/wp/2016/10/24/donald-trumps-chances-of-winning-are-approaching-zero/ and Dan Roberts, 'Donald Trump Lends Name to New Hotel so Near — and so far from — White House', *The Guardian*, 26 October 2016, https://www.theguardian.com/us-news/2016/oct/26/donald-trump-opens-international-hotel-campaign-trail-brand

9   See the Comment Section on Brett Sanders and Darren R. Reid 'Aftermath: A Portrait of a Nation Divided', *YouTube*, 11 October 2016, https://youtu.be/bU1wf4UIt-o.

There are certainly lessons to be learned. Context changes the meaning, and perhaps even the worth of a film. Prior to the election, the film was fairly criticised for not offering balance. In a post-election world, that imbalance (which had been forced on us by the silence of Trump's New York supporters) now appears to be the most important thing we could have captured. So much for the role of the filmmaker.

But if our authorial voice was challenged or altered by the electoral process, our role as lecturers was enhanced. Traditionally, the teaching of history, and more broadly that of the humanities, has involved the inculcation of critical thinking through the production, and criticism, of written texts. The assessment and dissemination of knowledge, and the demonstration of newly acquired skills of cognition, were primarily undertaken in a written form: essays, monographs, reviews, and so on.[10] However, with the democratisation of filmmaking technologies and the advent of smartphones with their increasingly capable cameras and powers of recording, historians, humanist scholars, and their students have been confronted with new challenges and opportunities. The usability of technology, its wider availability and mobility, allow new voices to be seen and heard in previously inaccessible spaces. The open-access nature of the online environment has destroyed previous barriers to distribution and dissemination.[11] The possibilities, and implications, for scholars are startling.[12]

Aftermath: A Portrait of a Nation Divided was an experiment in the pedagogic practices of humanists. It allowed us to involve our students in the creation of oral histories and the construction of the narrative that those sources informed. Our students were not the traditional

---

10 For a discussion of this issue, see David Theo Goldberg, *The Afterlife of the Humanities* (Irvine: University of California Humanities Research Institute, 2014), https://humafterlife.uchri.org/

11 Don Boyd, 'We are all Filmmakers Now — and the Smith Review Must Recognise That', *The Guardian*, 25 September 2011, https://www.theguardian.com/commentisfree/2011/sep/25/all-film-makers-smith-review

12 For a sample of the ways in which humanist scholars are utilising emerging technologies to challenge the traditional thesis, see Susan Schreibman, Ray Siemens, and John Unsworth (eds), *A New Companion to the Digital Humanities* (Chichester: John Wiley & Sons, 2016); Eileen Gardner and Ronald G. Musto, *The Digital Humanities* (Cambridge and New York: Cambridge University Press, 2015); David M. Berry (ed.), *Understanding Digital Humanities* (New York: Palgrave Macmillan, 2012).

synthesisers of content, but the producers of it — employing a transdisciplinary method in the disruption of a traditional subject.

As technologies evolve and change the way we live and communicate, it is imperative that post-digital-era graduates embrace new skills, and are capable of producing content across multiple platforms. On location in New York, our students were immersed in the making of history, learning to take the pulse of the city's electorate, collaborate with their lecturers, and shape the voices that informed the public debate. Understanding the language of film, and the rules that govern the interests and aesthetic preferences of the human eye were new avenues of discovery for our crew. Experiencing film production in Harlem, for instance, and engaging with its diverse community allowed our students to grow. They engaged with (and documented) the rich tapestry of that society; new technology was married with older methodologies. This was a digital humanist process in the sense that it was facilitated by new technologies, and it was post-digital in the sense that such technology serviced the pursuit of familiar intellectual and narrative goals.

In a post-truth world, humanities graduates must increasingly understand the construction of narrative, the 'truth' that permeates political and social cultures, and which defined the campaign of Donald Trump. In a year when opinion polls were found to be left wanting, failing to take account of a simmering nationwide desire for change, our film has become more relevant in the aftermath of Trump's unexpected victory. Instead of being a reassuring snapshot of a nation (un)divided, as it perhaps seemed to be when it was released, the film's inadvertent and renewed relevance stems from our failing to offer a voice to one side of the debate. Whilst the lack of balance initially drew criticism about our portrayal of New York's voters, in retrospect the silence of Trump's supporters in our film has become its most powerful feature — a deafening silence that changed the political landscape of the western world.

19. Post-Production Workflow

It is probably not too much of an exaggeration to say that documentaries are truly created during the post-production process. Of course, that could be said about most dramas as well, but documentaries are a particularly reactive type of film. Of all genres, they are most likely to be shot without the benefit of a script or pre-defined schema. Where a script does exist, the nature of the interviews captured, or the events documented, may well require the original structure, premise, or intellectual position be revised. Indeed, you must be open to change, minor or radical, throughout the production process. To be sure, it is entirely possible to construct a documentary film around a tight script which differs little from the final product, but even in those cases, the post-production process creates opportunities to change, innovate upon, or improve the original vision for the film.

The editing process presents filmmakers with a litany of possibilities. There is no one version of any single film, no inevitable final form that a production must take. The individual components of the documentary — contextual footage, interviews, animated sequences, voice-overs, connective tissue, soundscapes, music, and so on — can be combined in a practically infinite number of ways.[1] The same footage can be stacked, cut, juxtaposed, and remixed in such a mind-boggling variety of ways that the sheer number of possibilities can threaten to overwhelm you. At the start of the post-production process, then, you should take some time away from their footage. Moving straight from production to post-production (from shooting to editing) leaves little opportunity to recharge. In addition, some distance from your material will allow you to (re)appraise it from a fresh perspective.[2]

---

1   Sheila Curran Bernard, *Documentary Storytelling: Creative Non-Fiction on Screen. Fourth Edition* (New York: Focal Press, 2016), pp. 189–232.
2   Murch, *In the Blink of an Eye*, pp. 5–22.

Returning to your raw material, refreshed and reinvigorated, will allow you more easily to imagine the viewing experience you can create. To facilitate that process, you should ask yourself the following questions:

- What is the story (narrative structure) I want to tell?
- What is the most important story I can tell?
- What is the most important intellectual idea I can share?
- What are the key themes or ideas that my film needs to identify?

Working through these questions should help you to enter the post-production phase with a set of clear ideas and objectives. The answers to these questions may also highlight conflicting ideas that need to be resolved before your film can be constructed. Consider the subtle difference implied by the first two questions. Recognising and responding to this can be a challenge. But it can also be intellectually freeing and invigorating.

By the time you reach the editing phase in the production cycle, you will have likely been immersed in the creation of this work for weeks, if not months or years. Realising that an original concept may need to be revised or even abandoned may prove difficult, requiring you to excise significant amounts of prior work. If, however, you are able to recognise that there is a more compelling story to tell, or a deeper intellectual inquiry that can be made, it will almost certainly make for a superior final product.[3]

The second, third, and fourth questions are meant to encourage you to think about the ideas and themes that the post-production process can help you to realise. Are your preferred themes and ideas compatible with your initial vision; are the answers to those individual questions compatible with one another; have they changed over the process of your production? If, for instance, you find that the intellectual idea, your thesis, is no longer compatible with the narrative you believe your film should tell, it is likely that you will need to revise the intellectual basis of your work. This, in turn, will require you to revisit your film's structure and the key turning points faced by your audience and/or protagonist.

---

3   Sam Billinge, *The Practical Guide to Documentary Editing: Techniques for TV and Film* (New York: Routledge, 2017), pp. 190–97.

Nothing about your film is final until post-production is complete and your film is released.[4] You are not subject to your initial line of intellectual inquiry; as a result, you should be prepared for the possibility of further change and revision as the editing process proceeds. Remain flexible, in other words. Allow yourself to react to your footage. The following post-production workflow will allow you to work through the potentially daunting task ahead of you in a logical manner. Review your footage (all of it).

1. (Re)Consider your audience's relationship to the film.
2. Plan (or re-plan) a working structure for your film.
3. Begin creating a rough cut.
4. Step back from what you have produced.
5. Critically review the rough cut and reassess. If necessary, return to step two. Cut and replace those sections that do not work and preserve those that do. This process may involve a significant revision of your work. Once you have a rough cut that satisfies your intellectual criteria, proceed to the next step.
6. Begin refining your rough cut, paying attention to the timing and rhythm of the film.
7. Add polish to your film — colour-grade your footage, add music, adjust volume levels, add titles.
8. Step back from what you have produced.
9. Critically review your fine cut and reassess. If necessary, return to step seven and revise as necessary.

This ten-step process will help you to turn your raw, unedited footage into the best version of your film. Huge amounts of work and creativity will be involved in this process, probably at least as much as went into shooting and conceptualising your film. As a result, you should not be afraid to take your time in post-production. You should also be prepared for disappointment. There is every chance that sections of your film will not appeal to test audiences, requiring further work and revision.[5] This

---

4   Murch, *In the Blink of an Eye*, pp. 10–14.
5   Murch, *In the Blink of an Eye*, pp. 52–56; Hampe, *Making Documentary Films and Reality Videos*, pp. 307–08.

is, of course, all part of the process. Build disappointment (and the need to revise your work) into your expectations of what the post-production process will entail. Now, consider the post-production workflow in detail.

Review your Footage

Following the end of production, you need to acquire a firm grasp of all the footage you captured. You will be unaware of some of the successful (though unintentional) material that you captured, whilst some footage for which you had high hopes might, upon review, turn out to be unusable. It is, therefore, necessary for you to review every piece of footage you collected, taking detailed notes about what each video file contains. Unusable footage should be labelled as such, but notes should be taken as to why the footage is not usable — as the editing process commences, an 'unusable' shot may prove to have some use, albeit in an unexpected way. Every interview should be watched, from start to finish. Again, copious notes should be taken and, where possible, sections that directly speak to the main themes and ideas of your film should be carefully annotated.

Reviewing footage can be a tedious affair, often proving to be one of the least enjoyable aspects of the process, but it must be done fastidiously. The raw footage you captured represents the building blocks from which you will fashion your larger structure. Having an intimate knowledge of the footage you captured will allow you to begin envisioning the different forms your finished film might take.

(Re)Consider your Audience's Relationship to the Film

In many respects, discussions about structure are really discussions about how one might fashion a relationship between your film and your audience. As such, this stage in the post-production process should see you refining how you previously envisioned your project's structure.

Will your audience serve as the protagonist in a participatory experience (see chapter twenty-one); or will a more conventional on-screen protagonist or narrator be utilised instead?

Having reviewed your footage you may well find that your original plans are no longer suitable. Does the footage of your on-screen guide work as you envisioned it? If not, you may need to cut that idea and replace it with something else. By replacing an on-screen guide, however, the tone of your film — and the audience's relationship to it — may change substantially. This is something you will need to deal with in the next step of the process.

Plan a Working-Structure for your Film

Having reviewed your footage and considered the type of relationship you wish your audience to form with your piece, work can commence on the creation of a structure around which you will construct your film's rough cut; this is the point when your film will start to take on a meaningful shape. Until this part of the process, your documentary has been little more than an abstract, a collection of unconnected pieces of footage which could be assembled in any number combinations — a thoroughly theoretical proposition. When a structure is settled upon, something that resembles a film will begin to emerge from these building blocks.

When we started to assemble *Looking for Charlie*, we mapped out our working structure on three sheets of paper, each one representing an act of the film. With post-it notes and stills from our raw footage we then began to plot out a rough timeline, imagining the succession of sequences and ideas that our film would explore. At this early stage in the process, it was easy to over-stuff some sections whilst under-serving others. Post-it notes are easily amassed, and a design that appears to work on paper will not necessarily work when the editing process actually commences. Some sections will become dense and confusing whilst other will suffer from pacing issues and will require heavy revision. Still, this process allowed us to crystallise prior ideas whilst still experimenting with the different forms that the final piece could take.

The initial structure that you design should serve as a blueprint for your film — but be prepared to alter it, perhaps significantly, if the editing process demonstrates that parts of your plan are unsound. You must prepare yourself ahead of time to respond to your film as it begins to take shape. When some aspect of its emerging form does not work, do not be afraid to consider radical revisions. It can be difficult to set aside material that took significant effort and resources to film, but if it serves to create a more cohesive final product, such cuts or alterations should be embraced.

Begin Creating a Rough Cut

A 'rough cut' is the first draft of your film. During this process, the emphasis is not upon creating a releasable version of your film, but a version that is intellectually or emotionally competent. It is unlikely your rough cut will resemble a finished product, but it should at least be watchable to the filmmakers, if not to any outsiders. The creation of a rough cut should not see filmmakers overly concerned with precise matters of timing, of getting their edits exactly right. Nor should they be concerned with creating a cinematic look or feel through colour-grading and the precise organisation of music, and so on. Instead, they should focus upon the assembly process, of ordering shots and sequences to create a coherent narrative or an effective intellectual exploration of the subject at hand.

Raw footage should be combined with the other basic elements of the film. If a voice-over will be used, a working, rough commentary track should be recorded and added. You might record several versions of your voice-over — one deadpan, one conversational, and so on. Where necessary, sound from external sources (such as a lavaliere microphone and sound recorder) should be synced up with the appropriate footage. Complicated shots, which do not yet exist in a finished form — for example, an animation — should be represented by 'place holders'. These are typically simple blank screens with literal descriptions of the shots that will replace them. Some music, depending upon the importance it will play in the film, may be added to the rough cut. Final music selections (and timing) will be established at a later point in the process.

Despite being unwatchable to outsiders, the rough cut should provide the filmmaker with a reasonable idea of how their project, as it is currently designed, will turn out.

Step Back

By the time a finished rough cut is created, the likelihood is that you will have lost much of your objectivity — you will be so intimately connected to the material you have collected, and the rough cut that you have created, that you may find it almost impossible to assess it dispassionately. At this stage, therefore, you should consider taking a break from the process. Just as it is necessary to distance yourself from the project following the production phase, so too should the creation of a rough cut prompt another break. Only after you have been able to untangle yourself from the work will you be able to review the rough cut in a critical manner.

Critically Review and Reassess

Once you have gained some distance from your rough cut, you should arrange a private screening. To the extent that you are able, you should try to create an atmosphere that will allow you to appreciate the film as your intended audience will consume it. A projector would be ideal, but a large television in comfortable surroundings would also suffice. By moving away from the computer monitor on which the editing has been carried out, you will create a new contextual viewing experience which should allow you to achieve some degree of separation (and thus objectivity).

The screening of the rough cut should, as much as possible, occur organically; which is to say that you should avoid taking a significant number of notes as you watch it. Whilst there will no doubt be much to consider, all of that can wait for a second screening. In the first instance, you should attempt to keep this first screening as pure as possible. Most audience members will not be taking notes when they watch your film, so you should avoid this too. Instead, aim to open yourself up, as much as you can, to an organic viewing experience. Following this first review, the necessary post-mortem can begin.

Post-Mortem

Depending on how complex your project is, there will be much to analyse in your rough cut. It is not unusual for filmmakers to be deeply disappointed by the initial assembly of their material. Ideas that seemed to work perfectly on paper, or in the field, may not come together as expected. Finding yourself in such a situation, know that you are in good company.[6] Whatever issues you identify, they are likely to be surmountable challenges that the application of some imagination can repurpose into more effective sequences. If you find yourself uncertain, you should consider showing select moments from your work to trusted outsiders. They should be able to offer feedback on what does or does not work about a given sequence. If you find that you receive positive feedback about a section of your work that you find unsatisfying, it is likely that the issue is with the structure of your film — the sequence works, but not in context. Knowing this, you can reconsider and reappraise this aspect of your project and feed this perspective into the next edit.

Following this initial reappraisal, you should develop solutions to the issues you have identified. Are there problems with your narrative, or long sections that fail to engage? If so, consider new ways of presenting those aspects and implement them into a new rough cut of your film. Again, take some time away from your material and then reappraise it in another private screening. Repeat this process until you believe that you have a functional cut that is ready to be turned into a complete film.

Refine Your Rough Cut

Reviewing your rough cut should provide you with a clear sense of how your film is progressing and, in particular, its emerging strengths and weaknesses. When you are satisfied that it offers a solid foundation upon which you can build, you must then begin the process of refining it. This part of the process will produce a version of your work that will start to approach releasable quality. During the refinement process, you should pay particular attention to the timing and feel of your film. Shots

---

6   Empire of Dreams. Directed by Kevin Burns and Edith Becker. Los Angeles: 20th Century Fox, 2004.

or sequences that go on for too long, or are cut too abruptly, should be adjusted appropriately. All of your edits should be finalised so that the final pace of your film is realised.

Temporary music tracks should be swapped out for the music you intend to use, and working commentary tracks should be replaced with polished recordings. In addition to this, you should colour-grade your production, adding the final level of visual polish which will give your film a cinematic feel.

Step Back, Review Your Fine Cut and Reassess

You should now replicate the screening experience that you organised for your rough cut. Again, take some time away from the project in order to revisit the material with as much objectivity as possible. As you will be reviewing a near-final version of your film, you may wish to screen it with trusted friends or advisers. Naturally, however, they will be biased in your favour and keen to support you. Producing anonymised questionnaires to be completed after your screening may help to gather the type of critical notes from third-party viewers that you require.

Once you have screened your near-final cut, critically reassess its strengths and weaknesses and, if necessary, rework it to bring out the former whilst eliminating the latter. Add any final polish necessary to complete your project, including titles, any final editing decisions, and so forth. By this point, your production should now be complete.

Reflections

Throughout the post-production process, you should expect to be disappointed by your work as new cuts of your film emerge. This is completely normal and you should not unduly criticise yourself if your piece takes time to realise in the edit. You should also expect to be impressed with at least some of material you created. Post-production can be a time of significant highs and depressing lows.

You must also prepare yourself to solicit feedback and to respond appropriately. If you show a third party a rough cut, understand that they will not be able to fill in the blanks as easily as you. They will likely not understand the purpose of a rough cut and may, for instance, struggle to move past a lack of music, clumsy cuts, or poor editorial timing. Consequently, you should avoid screening rough cuts and instead solicit feedback only for material that is closer to completion. Criticism of your work can be difficult to process, particularly if you have invested significant time and resources into a project, but it is not the fault of your viewer if they do not enjoy what you have produced. Instead create a forum in which they can deliver honest, constructive feedback (such as through a questionnaire) which is directed and focused enough to help you as you continue the post-production process.

If you are prepared for the involved nature of post-production, you will be best positioned to take advantage of the many opportunities it offers you to realise the best possible form of your project.

20. The Three-Act Structure

Documentaries have more freedom to break the rules that dramas must typically obey. They tend to be self-aware and, as such, break the fourth wall. They often seem to lack traditional protagonists and antagonists and whilst some, such as Seth Gordon's *King of Kong* (2007), indulge this trope, many forego it. Despite all this, documentaries remain beholden to long-held structural expectations. A sound structure can help to turn any subject, no matter how seemingly banal, into an engaging intellectual experience. Likewise, any subject, no matter how inherently interesting, can be made uninteresting if it is explored in an unstructured or meandering manner. Facts and analysis may have significant intellectual value, but without attention to how audiences engage with (and absorb) cinematic formats, viewers can become lost or disinterested. You must, then, pay as much attention to the medium as you to do the message itself.[1] This is particularly true in the post-production process, when your film's structure is definitively realised. You may have had a sense of your work's structure early in the production, but it is during the editing phase that nebulous ideas are tested and the reality of your work becomes evident. Consideration of structure should therefore deeply inform this phase of your production.

The three-act structure creates a familiar and satisfying framework with which audiences are instinctively familiar. This allows filmmakers to set up a recognisable flow of information, which is easily consumed by audiences familiar and comfortable with this pattern. The presentation of the initial proposition and the first steps on the audience's journey occur in the first act; in act two, the substantive and most detailed part of the study is carried out; whilst in act three, the different intellectual or

---

[1] Richard Kilborn and John Izod, *An Introduction to Television Documentary* (Manchester: Manchester University Press, 1997), pp. 115–64.

narrative threads hitherto explored are brought to a clear conclusion. In other words, premise and context (act 1) give way to investigation and analysis (act 2) which, in turn, give way to reconciliation (act 3) of the different intellectual and narrative threads hitherto explored.[2] As in an academic paper, wholly new ideas should not be introduced in the third act; new information can be presented, of course, but this part of the film should instead focus on using that new information to resolve the ideas already established in the previous parts of the film.

The three acts should not be equal in length. Rather, the second act should be the most substantive component of the film, and the third act the shortest. Visualised, this is how the three-act structure might look for a feature-length documentary:

Fig. 56. The three acts of a production each has a distinctive role to play. The first act sets out the premise, core ideas, and principle argument (or line of inquiry) for the piece. The second act engages in the substantive investigation and analysis. The third act brings those core ideas and arguments to their fundamental conclusion.

In a short film, a similar structure can be employed. In an eight-minute film, for instance, a two-and-a-half-minute first act would precede a four-minute second act and a two-minute final act.

The second act, then, is the most involved portion of your work, the space in which the bulk of the intellectual exploration takes place. Setup and resolution (acts one and three) are just as important as what occurs in act two, but the uneven spread visualised above is a reflection of the need to focus these sections so that they appropriately prepare the viewer for, and pay off, the second act. A tight structure

---

2   John Yorke, *Into the Woods: How Stories Work and Why We Tell Them* (London: Penguin, 2013), pp. 24–31.

can significantly improve a project's 'watch-ability', and thus the ease with which audiences can engage with it.[3]

In specific terms:

Act one is about introductions and setting up a film's basic scenario. Who are the main players; what are their relationships; what are the questions, social needs, or external forces at play which will allow for an exploration of the main theme or topic you wish to analyse? In this act you must clearly identify the core element(s) that will unite the individual parts of your film, the project's intellectual through-line. Is it a particular individual's life; a question about a particular social or political experience; the exploration of a dominant idea or theme? If a documentary is about answering a specific question, the question should, in one form or another, be posed here alongside a rationale for why that question is important.[4]

Act two is when a film gets under the hood of its central conceptual mechanisms. In act one, the filmmaker introduces viewers to their intellectual world, setting up its basic rules, assumptions, questions, and so on. In act two they must then explore their core issues in depth.[5] In *Looking for Charlie*, a documentary about Charlie Chaplin, Buster Keaton, and the harsh realities of life in the silent film era, the first act set up a discussion about the ways in which contemporary society discarded performing artists who fell out of favour with audiences. In its second act, it makes the case that society is short-sighted because, even after performers have been discarded and forgotten, their influence is frequently long-lived. To facilitate the deepening of this discussion, the

---

3 Despite being a popular film, the ending of Peter Jackson's third *The Lord of the Rings* (2003) movie is often criticised. It seems to go on for too long — the audience keeps expecting it to end. From a narrative perspective, this extended ending allows for many emotional storylines to be resolved but, from a structural perspective, it is messy and unfocused, defying audience expectations to the frustration of some. For examples of some of the criticism of *The Return of the King*'s ending, see Jen Chaney, '"King" Gets Royal Treatment in Extended DVD', *The Washington Post*, 14 December 2004, and Andrew Blair, 'Ranking the Endings of *The Lord of the Rings: The Return of the King*', 8 September 2017, https://www.denofgeek.com/uk/movies/lord-of-the-rings-return-of-the-king/51754/ranking-the-endings-of-the-lord-of-the-rings-the-return-of-the-king
4 Yorke, *Into the Woods*, pp. 24–31.
5 Syd Field, *Screenplay: The Foundations of Screenwriting* (New York: Random House, 2005), pp. 89–105.

range of subjects in act two was increased substantially. The first act was primarily constructed around an exploration of the relationship between Charlie Chaplin and Marceline Orbes, the clown whose approach to pathos and comedy had so deeply inspired him. In act two, however, Buster Keaton and a range of other subjects, including the filmmakers themselves (in an autobiographical twist) were added to the mix. This growing cast allowed for overlapping experiences, perspectives, and themes to be brought to the fore; the case study in act one was thus transformed into the foundation for a discussion about the universality of the human experience in act two.[6]

Act three should then serve to bring the thematic and narrative threads developed in act two to a resolution. No new questions — at least major new questions — should be posed here.[7] In Michael Moore's *Capitalism: A Love Story* (2009), act three is the point when oppressed workers and other victims of the economic crash of 2008 are shown to begin a self-actualised recovery. Inspired by their actions, Moore then (literally) ties off the main themes of the film by sealing off Wall Street behind bright yellow 'crime scene' tape. Act three is when the beaten get back up, dust themselves off, and stare down the barrel in utter defiance. In the case of a factual documentary, this is the period at which truth, as understood by the filmmaker, is articulated in its clearest terms. Moore is melodramatic in his attempt to provoke his audience to action, but most documentaries end their films in a similar, though less on-the-nose, manner. The truth (or at least a reasonable candidate for the truth) has been revealed.[8]

Act two should have provided a deep enough exploration of the film's core issues that the conclusions generated in act three appear logical and justifiable. Indeed, the audience should receive a sense of intellectual (or, in the case of much of Moore's work, for example) emotional closure. Moore's ending to *Capitalism: A Love Story* is somewhat sentimental — in actuality, the actions of the workers are

---

6 *Looking for Charlie: Life and Death in the Silent Era*. Directed by Darren R. Reid and Brett Sanders. Coventry: Studio Académé, 2018.
7 Robert McKee, *Story: Substance, Structure, Style, and the Principles of Screenwriting* (New York: Harper Collins, 1997), pp. 303–16.
8 *Capitalism: A Love Story*. Directed by Michael Moore. Los Angeles: The Weinstein Company, 2009.

unlikely to have produced any serious, long-term improvements to their situation — nonetheless, their act of defiance, and the small victories they secure, leave the viewer satisfied. The film tells them that positive change can happen when people act to protect their own, collective interests.[9] This is a precise inversion of the film's opening sequence, which emphasised the powerlessness of ordinary people in the face of macro-economic forces. Moore thus brings his audience full circle on their intellectual and emotional journey, mirroring the film's opening portrait of despair with one of hope instead. One is, of course, free to disagree with Moore's thesis, but dismissing the effectiveness of his work is far more difficult.

In most examples, the third act of a documentary sees the filmmaker resolving their case. It is that resolution (even when it demands further action from the audience) that allows the film to end in a satisfying manner.

---

9 Ibid.

21. The Protagonist

Documentaries are journeys: frequently for a person represented on the screen, always for the audience. As such, the emergence of your film during post-production should be informed by a sensitivity to change. Subjects should be given room to grow and develop, should they require it. And your audience, likewise, should have opportunities to deepen their knowledge about a subject in unexpected but intellectually satisfying ways. Representing and guiding that growth can be challenge, but there are clear precedents available to you that can inform how you approach this aspect of your work.

Joseph Campbell argued that narrative is a vital part of the human perceptual experience.[1] It is in the details only that *Star Wars* (1977) is separated from *The Lord of the Rings* (2001–2003) and *Breaking Bad* (2008–2013). Walter White and Luke Skywalker might not appear to have much in common, but both *Breaking Bad* and *Star Wars* are about a character who a) craves change and b) through a shift of circumstances, is c) set on a path to realise some version of that change. Ultimately, both Skywalker and White are d) fundamentally altered by their quests to achieve some external goal, each becoming e) something the original character could not quite have envisaged at the start of their journey.[2] This narrative structure, in one form or another, is evident in a vast array of Western narratives. Documentaries, though ostensibly very different from dramatic films, are just as likely to utilise this journey as their fictive counterparts.

---

1 Joseph Campbell, *The Hero with a Thousand Faces. Third Edition* (New York: Pantheon Books, 1949; reprint, Novato: New World Library, 2008), pp. 1–40.
2 This breakdown of the protagonist structure is based upon Dan Harmon's 'Story Circle', which will discussed extensively in the next chapter. See Dan Harmon, 'Story Structure', Channel 101 Wiki, http://channel101.wikia.com/wiki/Story_Structure_101:_Super_Basic_Shit

It may be a false equivalence to talk about Walter White and Luke Skywalker in a discussion about documentary films, but consider Michael Moore's first film, *Roger and Me* (1989), in which the filmmaker attempts to confront General Motors CEO Roger Smith about the impact his company's downsizing policy has had upon Moore's hometown of Flint, Michigan.[3] In the film, Moore takes on the role of the film's protagonist and, just like Luke Skywalker and Walter White, he a) craves change (through confrontation) and so b) changes his circumstances (becoming a documentarian) so that he can set off on a quest c) to initiate the confrontation. Moore ultimately fails to force the confrontation with Smith but is nonetheless d) altered by the experience, learning much (which he communicates to his audience) throughout his journey. As a result, Moore e) finds victory in his failure, discovering a deeper truth despite his inability to achieve his original goal. Considered from a structural perspective, there is little that meaningfully separates Moore from Skywalker or White.[4] The substance of *Roger and Me* may be very different to that of a film like *Star Wars*, but the substructure of those films is remarkably similar. Even when no on-screen protagonist is identified in a documentary, one is always implied.

Consider Brian Cox's BBC documentary series, *Wonders of the Solar System* (2011).[5]

Viewers might reasonably assume that the series' charismatic presenter is its protagonist. This is not the case, however. Rather, it is the audience who unwittingly takes on that role and, in so doing, parallels the journeys taken by Moore, Skywalker, White, et al. It is, after all, the audience who a) craves a change in their initial state (to learn more) and, as a result, b) changes their intellectual circumstances by choosing to watch a documentary. From there they are able to c) confront their own ignorance, d) grow intellectually, face conceptual challenges, and e) emerge more enlightened.

Documentaries are, then, a form of participatory media. A distinction must therefore be drawn between those documentaries that feature an on-screen protagonist, like Moore in much of his work, and those that feature a guide whose principal responsibility is to

---

3   *Roger and Me*. Directed by Michael Moore. Burbank: Warner Bros., 1989.
4   Yorke, *Into the Woods*, pp. ix–xiv.
5   *Wonders of the Solar System*. London: BBC, 2010.

facilitate the audience's journey. Standardised narrative structures are common because they provide humans with a vector to understand the fundamentally disorganised and unstructured universe that surrounds them.[6] As a result, narrative provides you with a powerful tool. It can help you to construct texts that recognise the participatory nature of the viewing experience, whilst simultaneously shaping a production around the audience's role as active participants on an intellectual journey.

Harmon's Story Embryo

Campbell proposes a seventeen-point journey for the 'hero' protagonist. Producer and writer Dan Harmon (*Community* (2009–2014), *Rick and Morty* (2013-present)) offers a more streamlined version of this model which aspires to even greater universality — and which we will revise and refine for the documentary format. According to Harmon, most, if not all, successful narratives can be distilled down to just eight core elements, which can be found in virtually every compelling example of the form. Whilst it is certainly possible that Harmon may have overstated the universality of his case, the structure he proposes does fit a remarkable number of filmic narratives, fiction and non-fiction alike.

At the root of Harmon's argument is the idea that narrative, which he believes can be distilled down into a fundamental sub-structure he calls the story embryo, is hard-wired into the human imagination; that it serves as one of the key perceptual filters that allows the species to interpret and make sense of the world and their own lived experiences. As a result, fostering an accurate understanding of the universal mechanism of narrative, according to Harmon, has nothing to do with conforming to popular or transitory tropes or avoiding experimentation. Rather, it is an exercise in exploiting fundamental human psychology to create a method of information transmission which naturally resonates with an audience in an intuitive and impactful manner. It is, then, a tool that filmmakers can exploit to make their case in the most effective way possible.[7]

---

6   Stuart L. Brown, foreword to *The Heroes Journey: Joseph Campbell on his Life and Work* by Joseph Campbell (New York: New World Library, 2003), pp. vii–xii; Yorke, *Into the Woods*, pp. 33–34.

7   Dan Harmon, 'Story Structure', Channel 102 Wiki, http://channel101.wikia.com/wiki/Story_Structure_102:_Pure,_Boring_Theory

The story embryo argues that there are eight basic moments in any narrative which, together, make for an inherently satisfying structure. They are:

1. The coming of a **protagonist**.

2. That protagonist possesses a **need for change** (conversely, they may possess a particularly strong desire to maintain the status quo in the face of some external force).

3. The protagonist must then move beyond their status quo. They must **change their circumstances**; in other words, leaving their comfort zone.

4. The protagonist must then go on a **quest in search of what they desire**. If they wanted a change in their circumstances, they should attempt to realise that change. If they were taken out of their comfort zone by an external force, they might well be trying get back to their status quo.

5. The protagonist should then **find what they *think* they are looking for**. If they wanted an exciting life, they should now be immersed within it and, at some point, embrace that change.

6. The protagonist should then **suffer as a result** (undergo a setback of some kind).

7. **The protagonist must then recover** from point six, overcoming a setback they encountered in order to complete their narrative arc. In *Capitalism: A Love Story*, this is the point when the mistreated factory workers stand up for themselves against the corporate mechanisms that had hitherto exploited them.[8] In *Star Wars*, it is the point when Luke Skywalker resolves to join the rebel attack upon the Death Star, overcoming the death of his mentor, Obi Wan Kenobi.

8. The **protagonist** can then emerge from their recovery a changed, usually improved, person. The arc is complete.[9]

---

[8] *Capitalism: A Love Story*. Directed by Michael Moore. Los Angeles: The Weinstein Company, 2009.

[9] Dan Harmon, 'Story Structure', Channel 104 Wiki, http://channel101.wikia.com/wiki/Story_Structure_104:_The_Juicy_Details

In drama, the story embryo can be found in many films. The story of Luke Skywalker fits the model remarkably well, as does Michael Corleone in Francis Ford Coppola's *The Godfather* (1972), Woody Allen's Alvy Singer in *Annie Hall* (1979), Indiana Jones in Steven Spielberg's *Raiders of the Lost Arc* (1981), Jean-Pierre Jeunet's Amélie in *Amélie* (2001), and hundreds of others besides.[10] For filmmaker-scholars, this model is even more important when the audience's participatory role is recalled and utilised fully.

Casting the Audience as the Protagonist

When the audience is projected onto Harmon's model, no less than half of the protagonist's journey occurs before a single frame of film has been consumed. As the fulcrum in a participatory piece of media, the audience 1) is the protagonist, whose decision to engage with a documentary is 2) a product of their desire (or need) to learn more about a topic or perspective, and so, they 3) change their circumstances by placing themselves into a situation that will allow them to watch the documentary in question. This is part of the audience's 4) attempt to accomplish their goal — reach an increased state of enlightenment.

In this participatory model, the audience experience transitions into the hands of the filmmaker at the fifth point in Harmon's story embryo. The filmmaker, then, serves as a knowledgeable interlocutor, a guide, whose chief responsibility is to facilitate the final four stages in the audience's journey. In some documentaries, this role is filled in a rather literal way through the introduction of an on-screen guide — Brian Cox, Carl Sagan, Neil DeGrasse Tyson, and so on, serve as excellent examples. Such guides do not necessarily need to appear on-screen, however. They might only be presented as a disembodied voice (the narrator), speaking to the audience but never identifying themselves directly. Alternatively, they might not appear in any identifiable form whatsoever: a documentary with neither host nor narrator remains the product of its creator who, whether made manifest or not, remains the

---

10 *The Godfather*. Directed by Francis Ford Coppola. Hollywood: Paramount Pictures, 1972; *Annie Hall*. Directed by Woody Allen. Los Angeles: United Artists, 1977; *Raiders of the Lost Ark*. Directed by Steven Spielberg. Hollywood: Paramount Pictures, 1981; *Amélie*. Directed by Jean-Pierre Jeunet. UGC: Neuilly-sur-Seine, 2001.

audience's guide. As Alexander MacKendrick once put it, 'what a film director really directs is his audience's attention'.[11]

This is particularly true of the filmmaker-scholar, whose fundamental role is that of a guide. Because of this, points five to eight of Harmon's story embryo suggest you should not set out to guide the audience along a straightforward trajectory. Rather, you should first endeavour to lead the audience to a point where they 5) *think* they have found what they desire; enlightenment that superficially satisfies. In a documentary about the battles of the Second World War, for instance, an audience might reasonably expect, from an early stage, to have increased their knowledge about the mechanics and tactics of battle. The audience should thus have this desire validated by the filmmaker.

However, the documentary should then seek to 6) problematise the audience's expectations by presenting a deeper intellectual experience than the audience could have anticipated at the outset. After a discussion about battlefield tactics, the documentary might then begin to explore the human cost of conflict; this point in the film, then, should open the audience up to new intellectual possibilities beyond those they initially imagined when they first engaged with the piece. This ever-deepening intellectual discourse ultimately 7) resolves the problematisation of the previous point; the acquisition of deeper and more sophisticated knowledge or modes of thinking should come to self-evidently justify the unimagined places the filmmaker has taken the audience. By the end of the film, the audience 8) should exit the process changed. Not only has your film helped the audience to increase their store of knowledge, as they had originally hoped, it should also have increased their understanding of the subject in ways they had not previously anticipated.

A poorly constructed documentary is one that fails to challenge its audience. This would, according to Harmon's model, vastly reduce a film's ability to impact the viewer. As such, point six, the intellectual pivot, should be of great structural importance to you.

In *Wonder of the Universe* (2011), the *challenge moment* occurs when Brian Cox addresses the inevitability of the universe's end. The philosophical questions raised by this moment, and the implications for the value we

---

11   Alexander Mackendrick, *On Filmmaking* (London: Faber & Faber, 2006), p. 200.

attach to life, are potentially astounding. Cox, however, reassures his audience through a follow-up discussion: a doomed universe is still a marvel, even if its end can be predicted. That something reaches a conclusion, Cox suggests, does not reduce its beauty or significance[12]. In Harmon's parlance, the audience suffers, they recover, and exit the film in a changed state (more enlightened).

Of course, point six in this model (the problematising pivot) should not replace a clear statement of intent (or thesis) presented at the outset of a documentary. As with an academic paper or monograph, the point of a film should be clear to the audience from an early stage. Point six, however, should serve as the moment at which some unexpected depth, or intellectual inquiry required to prove that thesis, should occur. The following discussion (point seven), should then serve as a form of intellectual reconciliation; enlightenment should follow problematisation. The thesis of a given documentary may, in its own right, offer surprises or challenge conventional wisdom, but Harmon's story embryo requires a deeper intellectual pivot, needed to prove an already disruptive thesis, which will set the stage for a keystone discussion.

Harmon's story embryo essentially streamlines Joseph Campbell's 'Hero's Journey'. When used in relation to the documentary, however, it suggests that half of the experience is controlled directly by the audience. Whilst the audience is vital in any form of participatory media, this does create a misleading impression about the balance between the agency of the filmmaker and the audience. As a result, a further refinement — the documentary embryo — is required to describe documentary structure more accurately:

1. By watching a documentary film, the audience makes the decision to embark on a quest towards enlightenment and so initiates a **participatory** experience (watching a documentary).

2. On that quest they meet a guide (the filmmaker or their proxy) who helps them to **discover** the types of **information they expected to learn**.

---

12 *Wonders of the Universe*. London: BBC, 2011.

3. A deeper intellectual process then reveals **new information, or a new perspective** which complicates the audience's view of the subject.

4. That complication is then **intellectually resolved**, and the audience's understanding is thus deepened in a way they might not have expected at the outset.

5. The intellectual process is then brought to a close, reconciling the audience's pre-existing perspective with the knowledge they have newly acquired. The film's principal ideas are brought to a **conclusion**, which leaves the audience satisfied that their quest was not only worthwhile but deeper than they anticipated.

Superimposed onto a three-act structure, the participatory documentary structure can be visualised thus:

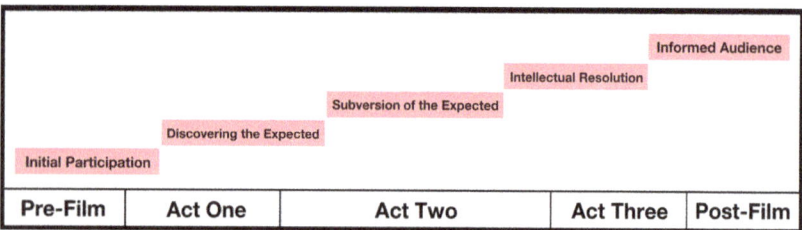

Fig. 57. The documentary embryo overlaid onto the three act structure.

Of course, rules (and structural models) can be challenged. Before disregarding the documentary embryo, however, we would encourage you to consider seriously the logic of its structure. Breaking rules can have positive results, but they can leave viewers disorientated and, if not handled well, disgruntled. Mark Cousins' experimental documentary *Atomic: Living in Dread and Promise* (2015) offers neither an on-screen guide nor a narrator, a reality that is complicated by only a small amount of incidental dialogue which does not articulate a clear message or narrative. In spite of this, its problematising pivot is clear and satisfying: after significant immersion in the horrors of the atomic age, images of MRI machines and other peaceful, constructive uses of nuclear technology, challenge the viewer. The result is a film that underlines the dangers of nuclear technology even as it acknowledges the good that can come from it. Horror is thus tinged, as the film's subtitle promises, with promise. The complexity of the nuclear question is therefore established in the minds of the audience, as the three-act structure collides with a participatory model of audience engagement.[13]

The On-Screen Protagonist — The Journey

Whilst the audience can certainly serve as an abstract model for the protagonist, there are more conventional opportunities to apply character-driven narrative models to the medium. By building a documentary around the experiences of an individual (or small group), be they the filmmaker or a third party, an on-screen protagonist will naturally emerge. In the case of a third-party subject, such as a historic or contemporary figure, narrative models rooted in Campbell's 'Hero's Journey' and Harmon's story embryo prove to be particularly useful.

In Banksy's *Exit through the Gift Shop* (2010), a protagonist-centred structure allows for the commercialisation of the street-art movement to be explored through biography. In the film, Thierry Guetta is 1) identified early-on as the film's protagonist. He has 2) a desire to make a valuable contribution to the street-art community. As a result, he 3) reinvents himself to become its principal documentarian, 4) pursuing the ever-elusive Banksy to ensure that he captures a complete record of the movement's most important figures. Over time, 5) Guetta and Banksy develop a friendship which leads the artist to invite Guetta to produce a documentary about the movement, but, as Banksy discovers, 6) Guetta was woefully incapable of creating a watchable film and, as a result, Banksy sidelines him from the project. Responding to Banksy's suggestion that he produce some art of his own, Guetta (7) hatches a plan to become a self-made street-art phenomenon. In spite of a lack of artistic skill, he uses his connections in the field to launch his new career and, in the process (8) reinvents himself. By the end of the film, Guetta has graduated from filmmaker to a leading light in the field he once documented; his unsuitability for either role serves a warning about the thin line that can separate hype from substance.[14] Like so many

---

13   *Atomic: Living in Dread and Promise*. Directed by Mark Cousins. London: BBC, 2015.
14   *Exit through the Gift Shop*. Directed by Banksy. London: Revolver Entertainment, 2010.

dramatic films, *Exit through the Gift Shop* relies heavily upon a familiar protagonist-centric narrative.

By employing a familiar narrative structure that hits each of the major pivots described by Harmon's story embryo, Banksy no doubt over-simplified much about Guetta's life, but the result is a compelling narrative which allowed for the pursuit of a deeper truth about the commercialisation of street art. Still, ethical questions abound, not the least of which is the extent to which filmmakers should bend or shape their subjects to fit a pre-determined structure. The answer to this quandary is simple: if a subject's life does not fit a recognised narrative model (and, therefore, is unlikely to contain the tensions and narrative shifts that will arrest an audience's interest), they should not be employed as a protagonist. In other words, do not make your subjects fit a structure for which their lived experiences are ill-suited. When a filmic structure fails to enhance one's analysis of a subject, a different approach should be taken. Appealing to the documentary embryo, and centring a film on the audience, may suffice but in cases where a single subject (or small group) sits at the heart of a film, audiences might well expect that subject to be explored in a familiar way.

In such instances, the filmmaker (or a proxy, acting on their behalf) might serve as a suitable protagonist around which a familiar and engaging structure can be woven, which intersects with the chosen subject. Journeys of intellectual discovery are common, with on-screen hosts taking their audiences on journeys centred on personal quests of discovery or self-improvement.

'The Journey' is common in a wide variety of documentaries. Indeed, it is so common that it is often used in trite, unimaginative ways: after identifying 1) themselves as the film's protagonist and 2) articulating their desire to learn about subject X, the on-screen host can 3) move out of their traditional lives in order to start a journey of 4) discovery about the subject at hand. Along the way they will 5) start to achieve their goal, learning much, but they will 6) also discover unexpected truths. Ultimately, however, they will 7) reconcile those discoveries with their pre-existing expectations to arrive at a new truth and, consequently, 8) leave the process with a deeper understanding of their subject.

Consider the above abstraction and compare it to any number of broadcast documentaries, particularly those in which a non-expert, typically a celebrity of some kind, goes on a journey of discovery, perhaps to uncover the truth of their family history. In many cases, this structure is used in poor-quality or mediocre documentaries, but the device itself serves to effectively dramatize events and studies which, otherwise, might fail to retain the interest of a broad audience. But any structure is only as valuable as its implementation, and whilst there are innumerable examples of 'The Journey' that are derivative, unimaginative, and uninteresting, these are problems with individual productions, not necessarily the structure itself.

'The Journey' needs to be a narrative that is worth telling in its own right. Authenticity and honesty are vital to the successful use of this device, and genuine autobiography, which brings out deeper themes in a study, can add compelling new insights to an intellectual discourse. Broadcast documentaries in which on-screen hosts stage aspects of their journey for the sake of creating a narrative can alienate discerning viewers. More effective than a staged and dishonest journey would be a complete reappraisal of how the rules of cinematic narrative can best be used to engage an audience with the subject at hand.

Structural models must be used in imaginative and appropriate ways to pursue a deeper, more meaningful discourse. 'The Journey' is an excellent example of a documentary trope that has been overused in derivative ways. British documentarian Louis Theroux has used it throughout his career to varying degree of success. In *My Scientology Movie* (2015) he succeeds to a greater degree than he does in many (though certainly not all) of his prior productions. Because Theroux is documenting a group in whom he has a genuine interest and in whose religion he has a solid intellectual grounding, his journey in that film feels real. The result is a high-quality production in which Theroux's growing discomfort carries significant intellectual weight. The audience is able to believe that Theroux is going through a (re)formative process.[15]

The three-act structure, story embryo, and its derivative, the documentary embryo, are devices that are only as effective as their implementation. Utilising a structural model does not guarantee that an effective film will be produced, though it may increase the likelihood that this will occur. Likewise, disregarding such structures will not

---

15 *My Scientology Movie*. Digital Stream. Directed by John Dower. London: BBC Films, 2015.

necessarily lead to a poor-quality product; nonetheless, thoroughly understanding the structures or narrative conventions most audiences expect (and even demand) will make it easier to challenge dominant narrative models in the documentary space.

22. Assembly

With a clear sense of how you intend to structure your film, the actual assembly may feel like a formality. But the construction phase is not merely a technical exercise; significant creative freedom exists, even if you now have a well-developed schema. The ways in which sounds are layered, the choice of music, the types of cuts of you utilise — all will help to shape the intellectual and emotional impact of your work.

To be sure, a degree of technical expertise is required for this phase of your project. If you have a collaborator who possesses the relevant editing skills, it may be appropriate to leave the technical side to them. If that is not the case, however, understand that, just as with the process of learning how to capture footage, the basics of editing can be learned quickly, whilst practice and dedication will deepen your skills over time. The assembly phase is less about technical skill than it is creativity and experimentation. There are three processes that will allow you to continue to add depth to your work: editing, colour-grading, and sound-tracking.

Editing

The most important part of the post-production process, editing, transforms raw footage into a cohesive whole, but it is much more than that in practice. The individual units of cinematic language — shots, sequences, music, soundscapes — need to be assembled into an accessible audio-visual dialogue, the on-screen equivalent of sentences, paragraphs, and chapters. Whilst much has been written about the editing process, from both a theoretical and practical perspective, the power of visual grammars comes from their versatility, their ability to reflect the ideologies and mental processes of the filmmaker (and of their audience). In other words, every film defines the contours of its own visual syntax, setting parameters of understanding and interpretation which, within the film's own context, can be built upon or, as necessary, defied. These grammars, in turn, speak to a much larger body of filmic works, the overall language of film, within which you must define your own dialect and accent.[1]

Fig. 58. The Odessa Steps sequence. *Battleship Potemkin* (1925). Directed by Sergei Eisenstein (0:48:15–0:56:03).

According to the legendary Soviet filmmaker Sergei Eisenstein, the true power of film is not to be found in any individual shot; rather, it comes from the juxtaposition of different images as they are presented sequentially. Eisenstein called this the 'montage' and, to him, it was one of the most powerful, fundamental devices available to filmmakers. To layer images in sequence was, Eisenstein posited, to layer them vertically in the audience's imagination and, in so doing, to engage in a profound act of creation. Alongside Vsevolod Pudovkin and Dziga Vertov, Eisenstein emphasised the raw power of the editing process, its ability to create tension and to stir emotions in one's audience. His epic *Battleship Potemkin* (1925), with its famous Odessa Steps sequence (which would form the basis of a similar scene in Brian De Palma's 1987 film, *The Untouchables*) is a demonstration of the power of effective editing.

---

1 Mackendrick, *On Filmmaking*, pp. 3–35.

No inter-titles, dialogue, or music are required to communicate the emotions and horror of the Odessa Steps. Happy spectators wave. We see images of smiling faces, the young, and the elderly. Suddenly, the people begin to run, charging down the steps as looks of adulation turn to horror. We see images of the military advancing. Bodies begin to collapse upon the steps. The military continues its advance. Shots are fired. We see close-ups of terrified faces; a wide shot of the fleeing masses; close-ups again, as looks of fear and confusion abound. The crowds continue their flight down the steps. A child falls, his mother, unaware, keeps running. We see a close-up of the child's bewildered face. The mother stops and slowly looks back. A close-up of her face; suddenly horror and realisation spread across it. The editor cuts back to the child, blood dripping down his forehead. He is screaming and reaching out towards the camera. He passes out. We cut to the mother, her face now a mask of existential dread. Cut to the boy, unconscious, with feet and legs surrounding him as those who are fleeing pass around and over him. We see an extreme close-up of the mother's eyes, wild terror engulfing them. The surge of the masses intensifies. There are close-ups of walking canes and feet landing upon the boy's prone body. The editor cycles through images: the mother's anguished face; her son being trampled; wide shots of the masses fleeing; the mother's anguished face; the boy's body; the mother's anguished face. The cuts, like the impacts to the boy, come quickly.

It is a devastatingly effective sequence, its potency undimmed by the passage of time. The power of these edits cuts across generational and cultural divides, speaking to audiences as clearly in the 2020s as it did in the 1920s. The power of the edit is supremely showcased by this sequence.[2]

The power of editing fascinated early Soviet filmmakers, partly because conditions in the USSR following the Bolshevik Revolution, where celluloid was available in only limited supply, necessitated short takes and their imaginative assembly.[3] But this fascination only hinted at the editing process's versatility. Imaginative assembly can lead to stirring results, and inspiration need not be sought in theoretical texts

---

2 Dancyger, *The Technique of Film and Video Editing*, pp. 13–26; *Battleship Potemkin*. Digital Stream. Directed by Serge Eisenstein. Moscow: Goskino, 1925.

3 Rhode, *A History of Cinema from its Origins to 1970*, pp. 79–116.

alone. You should expose yourself to a variety of different cinematic dialects prior to editing and reflect deeply on the edits you see.

In *The 39 Steps* (1935), Alfred Hitchcock cuts from an image of a woman screaming to an image of a steam train rushing towards the camera. The shots are unified by a common sound, the screech of the train's whistle. The whistle abstractly replaces the sound of the woman's terror before, moments later, finding a more literal purpose alongside the image of the approaching steam train. This imaginative cut underlined a connection that the visuals had already helped to establish; it complemented and enhanced them, allowing the filmmaker to make his point in a more emphatic, and chilling, manner. Such asynchronous cuts can help to build tension or deepen the sense that events overlap, or are somehow connected, as sound from one part of a film bleeds through to another. It is a subversion of a reality, which can, if used appropriately, help to deepen the audience's immersion in your work.[4]

Just as striking, though for different reasons, is the match cut. In David Lean's *Lawrence of Arabia* (1962), Peter O'Toole's T. E. Lawrence spends a few moments staring at a lit match as it burns towards his fingers. He blows it out. Cut to a shot of the desert, the sky bleached red as the barest tip of the sun emerges from behind the horizon. As one light goes out another, very different form of light, utterly beyond the control of human beings, comes into being. An ending and a beginning, interior to exterior, the controlled and the uncontrollable. The shots mirror each other, symbolic opposites but physical parallels. The result is deeply effective.[5]

In both *The 39 Steps* and *Lawrence of Arabia*, a non-verbal connection between different events and locations is made through the power of the edit. The individual shots that make up each of these cuts are effective in their own right, but together they create a more powerful whole; a combination of symbolism and abstract depth, which helps to enlighten the audience without having to directly, or bluntly, tell them the desired information. In much the same way, you should aspire to make cuts that successfully deepen your audience's understanding of the issues at hand. Neither abstract symbolism nor Hitchcockian levels of innovation are strictly necessary, only a focus upon utilising each and every cut in

---

4 Dancyger, *The Technique of Film and Video Editing*, pp. 88–90.
5 Gary Crowdus, 'The Editing of Lawrence of Arabia', *Cinéaste* 34 (2009), 48–53.

the most effective way possible. Careful review of precedent, with an eye trained upon the ways other filmmakers have handled cuts between and within sequences, will pay intellectual dividends.

If an interview is filmed using three cameras, each resultant angle should serve a different communicative purpose: a wide shot might show the subject in context; a mid-shot might serve to bring the audience within a relatable distance of the subject; whilst a close-up might reveal new levels of emotional truth by focusing the viewer's attention upon otherwise indiscernible changes in the interviewee's facial expressions. Cutting between these three angles should not, however, be an arbitrary exercise. Rather, each cut should be used to reflect or counterpoint some detail in the subject's testimony. Cutting from a mid-shot to a close-up could, for example, help to underline a change in the facial expression of your subject. Should the subject then withdraw into themselves, offering more limited access to their emotional world, it would make logical sense to cut back to the mid-shot. This cutting sequence (mid-close-mid) should help to draw the audience's attention to this change.[6]

The editor can also play with time, drawing out moments or streamlining them to achieve noticeably different effects. By cutting from one part of a shot to another (without changing to another camera angle) in a single sequence, the editor will create a noticeable jump as a scene moves from one state to another without showing the intervening steps. By cutting from point A to point C, you can draw attention to the absence of B.[7]

Such jump cuts can be used to communicate anxiety or to help build tension. In *Roger Waters: The Wall* (2014), jump cuts were used extensively in the film's early sequences. The film documents the journey of former Pink Floyd front man, Roger Waters, as he explores the thematic roots of the band's 1979 opus, *The Wall* (1979), by interspersing an autobiographical, reflective journey about the nature of war with live concert footage. In the opening sequences, a multi-camera setup allows Evans to film Waters from numerous angles as he prepares to begin a deeply personal journey of discovery. Jump cuts add a sense of uneasiness to the sequence, as if much has been left unsaid. The sheer number of these cuts draws attention to the mundane nature of Waters'

---

6   Mackendrick, *On Filmmaking*, pp. 251–71.
7   Billinge, *Editing*, pp. 218–32.

preparation, whilst simultaneously giving it weight. Time becomes difficult to measure when jump cuts are employed. Have a few seconds been removed, or have entire hours been excised from the process? Perhaps, these cuts imply, time does not matter at all.[8]

Editing, then, is not a mechanical process, but a deeply creative one. Cuts within and between sequences can create meaningful depth, which enhances raw videography. Cutting from a person's face and upper body to a shot of their hands might provide the audience with additional insight into a person's inner emotional state. Drumming a distracted rhythm on one's thigh or the clenching of fists can communicate a lot of information that might otherwise go uncommunicated. The timing of these shots, the duration for which they linger on screen, their relationship to the next image in the sequence, all can create a powerful impression in the imagination of the audience.

As editor, you will have many tools at your disposal. Some of the most important are:

- **Hard cut**: cutting from one sequence to another without a transition. A very common edit.

- **Match cut:** just as the match going out cut to the rising sun in *Lawrence of Arabia*, match shots combine moments that mirror or invert one another. They are cuts between images that are symbolically related but physically distinct.

- **Asynchronous sound cut:** the sound from one shot bleeds into another.

- **Parallel editing:** explore parallel events by cutting between them in the space of a single sequence. Using parallel editing allows you to compare or contrast concurrent streams of imagery or contrasting phenomena.

- **Cutaway:** cut from the main focus of a sequence to a detail, such as a cut to the fidgeting hands of an interview subject or the object at which they appear to be staring before cutting back to your principal subject.

---

8   *Roger Waters: The Wall.* Directed by Sean Evans and Roger Waters. Universal City: Universal Pictures, 2014.

Colour-Grading

The colour-grading process can also be used to deepen a film's visual subtext. A more cinematic feel (unnoticed, but appreciated by audiences) can be achieved by using features in your chosen editing software that will allow you to control the shadow and highlight levels of your footage in order to emulate the effect of shooting on celluloid. By deepening the shadows and increasing the vibrancy of highlights, you will broaden the perceived colour range of your footage by creating a greater contrast between the light and dark areas in your frame. Software such as Da Vinci Resolve or Adobe After Effects can provide significant control over the colour palette of your film whilst apps such as iMovie on the iOS allow for basic colour-grading to be carried out on a tablet and mobile device (see video lesson ten, located in chapter twenty-three).[9]

Aside from emulating the look and feel of celluloid, colour-grading can be used to code meaning into your films more substantially. The saturation level of your sequences, for instance, can be increased, to give your footage a richer sense of colour, or decreased in order to give it a bleaker, washed-out tone. Greater levels of colour might reflect a sequence in which vibrancy is an important theme, whereas a washed-out, desaturated sequence might more effectively convey a less optimistic subtext.[10] During the post-production process for *Aftermath*, we desaturated much of our footage in order to underline the pessimistic outlook most of our subjects envisioned under a Trump presidency.

In *Looking for Charlie*, we removed all colour and instead graded for a celluloid-like black-and-white look. As a film about silent cinema, it made perfect sense for us to develop such an aesthetic, but it was practical necessity that encouraged us to embrace this fully. Because were using a mixture of cameras, some of which captured a broad dynamic range (a wide colour spectrum) and some which did not, creating a cohesive look between different shots proved difficult. By removing all colour from

---

9   Dion Scoppettuolo and Paul Saccone, *The Definitive Guide to Da Vinci Resolve* (Blackmagic Design: Port Melbourne, 2018), pp. 287–366; Mark Christiansen, *Adobe After Effects CC: Visual Effects and Compositing Studio Techniques* (Adobe: New York, 2014), pp. 197–202; Tom Wolsky, *From iMovie to Final Cut Pro X: Making the Creative Leap* (Focal Press, New York, 2017), pp. 285–314.

10 Alexis Van Hurkman, *Color Correction Handbook* (New York: Peachpit Press, 2014), pp. 83–113.

our footage, and grading for a consistent black-and-white contrast ratio, we were able successfully to match footage produced by very different cameras. The theming of the documentary complemented this aesthetic choice, as did our decision to release it as an exhibition film, screening it in venues related to silent-era film and cinema history. A prestige-style black-and-white aesthetic perfectly reflected the subject and era covered by the film, and the spaces in which it was shown.

Fig. 59. A still from one of the earliest films. The difference between the highlights (light areas) and shadows (dark areas) captured by celluloid are stark and evident here. This effect can be emulated by deepening shadows and blowing out highlights in post-production software. *Train Pulling into a Station* (1895), directed by Auguste and Louis Lumière.

Colour levels should be consistent in any given scene — sudden changes can distract audiences and break their immersion in your work. Beyond the individual scene, however, you should feel comfortable in altering colour palettes to suit the needs of a given sequence. Some sections of a film may, for instance, employ a desaturated palette whilst, in others, the saturation level may be increased. Such variances in colour profiles should not be arbitrary, however. They should reflect tonal, thematic, or chronological shifts in your narrative. Just as altering aspect ratios can recall ideas about classic or modern cinema, so too can different colour profiles be used to differentiate one part of your work from another. For example, you might stylise re-enactment to give it a vintage feel, whilst leaving modern interview scenes largely untouched. Such variable colour palettes can be used subtly to colour-code your film, to help the audience keep track of their temporal location within the narrative.[11]

Colour-correction software can also be used to fix issues that were baked into the footage as it was captured. Basic settings in your chosen software, such as exposure, brightness, and contrast, can be used to modify footage that is, in some way, in need of correction. If you over-exposed your footage, for instance, using a combination of the exposure and brightness functions in your chosen software package should help you to reduce the impact of this error. Be aware, however, that only so much can be accomplished in post-production; minor errors can be corrected, but more significant issues will require that you reshoot the scene entirely.

As with editing, successful colour-grading is a process that requires practice. The basics are comparatively easy to grasp, but mastery will only come with experience. Colour-grading should occur in the following three phases:

- Correct any necessary errors in your material, such as over-exposure, using basic software features such as exposure, brightness, and contrast controls.

- If desired, grade your footage to emulate the feel of celluloid (deepen shadows and blow out highlights to increase perceived colour depth).

- Stylise your footage using the more advanced tools in your software package or app.

To introduce you to the colour-grading process, we have prepared a video lesson that will teach you core techniques in Adobe After Effects (see chapter twenty-three)

Sound-Tracking

Whilst effective editing (and colour-grading) can do much to create an immersive filmic experience, the music that you employ can add additional depth to the audio-visual experience. Whether used sincerely

---

11 Alexis Van Hurkman, *Color Correction Look Book: Creative Grading Techniques for Film and Video* (New York: Peachpit Press, 2014).

or ironically, music can serve as a reflection or counterpoint to the visual aspect of your film, allowing you to add another layer to engage and entertain your audience.[12]

Whatever one thinks of his political stance, Michael Moore's use of music in his films is frequently effective. Often ironic and unexpected, Moore's use of music, like that of Quentin Tarantino, adds layers of sincerity, irony, and style to his work. At times, Moore uses music sincerely, to help evoke a specific emotion in his audience, as he did in *Capitalism: A Love Story* with the Irish folksong, 'The Last Rose of Summer' (1805). In a scene near the end of the film (1:57:20–2:00:21), Moore speaks in solemn tones about the death of Franklin D. Roosevelt and the country's subsequent move away from economic progressivism. As he does so, the opening chords of the song play. When he finishes his speaking, the music swells over footage of Roosevelt's funeral. After a short break, Moore's commentary resumes and he lists all of the rights that Roosevelt had envisaged but were not enacted. In the last part of the sequence, the song continuing to play, Moore shows footage of the aftermath of Hurricane Katrina, the natural disaster that devastated communities across southern parts of the US and, specifically (and most famously), in New Orleans.[13] The last rose (Roosevelt) was dead, and the summer (political support for workers) was at an end. It was not a particularly subtle moment, but it was effective.[14]

In contrast, Moore's use of The Go-Go's 'Vacation' (1982), an upbeat pop song, in *Fahrenheit 9/11* over footage of George W. Bush golfing as American troops were being deployed in the Middle East, was deeply ironic. In that section of the film, 'Vacation' underlines the apparent frivolity of the president's life compared with the vast responsibilities he was, according to Moore, actively avoiding.[15] The use of Richard Hawley's 'Tonight the Streets are Ours' (2007) in Banksy's *Exit through the Gift Shop*, an upbeat, retro-style track, similarly helped that filmmaker

---

12   Andy Hill, *Scoring the Screen: The Secret Language of Film Music* (Milwaukee: Hal Leonard Books, 2017).
13   For a discussion on the social and cultural impact of Katrina see Jean Ait Belkhir and Christiane Charlemaine 'Race, Gender, and Class Lessons from Hurricane Katrina' *Race, Class and Gender*, 14: 1/2 (2007), 120–52.
14   *Capitalism: A Love Story*. Directed by Michael Moore. Los Angeles: The Weinstein Company, 2009.
15   *Fahrenheit 9/11*. Directed by Michael Moore. Santa Monica: Lionsgate, 2004.

to set a suitably irreverent tone for his work.[16] Footage of street artists being chased by the police stands in contrast to the upbeat melodies of Hawley's music, hinting at some of the deeper themes Banksy hoped to explore. It was an absurd, entertaining piece of foreshadowing and irony that worked extremely well in context.

Of course, it is difficult, if not impossible, for independent filmmakers to secure the necessary rights to include popular music in their work. The costs are outrageously prohibitive. Rather than thinking in terms of pop music, think instead in terms of mood and tone. Popular artists may be out of reach, but viable alternatives are available. A plethora of royalty-free recordings, covering a vast array of genres, are released every year by relatively unknown artists, some of which are of an extremely high quality. Royalty-free music tends to require the purchase of a license, resulting in an up-front cost but, particularly for budget-minded filmmakers, there are some royalty-free collections that do not require an upfront payment of this nature. Examples include Musopen.org (an excellent source of public domain recordings of classical music), The Free Music Archive (a mix of free and paid-for music and songs) and Premium Beat (paid-for music). Significant time and effort will be needed, however, to find material suitable for your project. Royalty-free music varies in quality and suitability and you may need to listen to hundreds of tracks before finding a suitable addition to your sound-track. When that discovery is made, however, the effect can be tremendous. Whatever music you select, use it imaginatively and with care. Even high-quality music can be used in ineffectively.

The use of music should be varied and considered. It can add to background ambience, help to sincerely appeal to the audience's emotional state, or make bold ironic statements. The creative potential it offers you is substantial.

Beyond royalty-free collections, bespoke music can be commissioned. Whilst not always cheap — and certainly not a guarantee of quality — websites and online spaces that specialise in the hiring of people with creative skillsets will allow you to engage with musicians and composers of varying skill levels. In *Looking for Charlie*, we utilised this option extensively, commissioning two pianists to produce a range

---

16 *Exit Through the Gift Shop*. Directed by Banksy. London: Revolver Entertainment, 2010.

of instrumental tracks. Some of these were original compositions, whilst others were new versions of copyright-free music from the nineteenth century. Our most audacious commission for the film was a three-track jazz drum sound-track recorded by a Parisian musician. Combined with other, royalty-free sources of music, this provided us with a varied and effective soundscape, which we employed extensively through the film.

As with every other aspect of the assembly process, we encourage you to embrace the opportunities offered when you are constructing your sound-track. It is one of the final opportunities you will have to craft and shape your audience's journey.

23. Editing Workflow in Adobe Premiere Pro

Fig. 60 Watch this video lesson for an in-depth introduction to editing in Adobe Premiere Pro. https://hdl.handle.net/20.500.12434/6ff71a81

There are many different pieces of editing software available, ranging from powerful but free (or low-cost) apps, to more versatile packages which bring with them a more significant financial outlay. For the purposes of this chapter, we have chosen to provide a walk-through of Adobe Premiere Pro. It is powerful and an industry standard. Though it is not free, it is available as part of a competitively priced monthly subscription which should place it within the means of many readers. Alternative software packages are available, many at a lower cost with a similar set of features and workflow as that employed by Adobe's software. Even if you choose to utilise a different software package, the basic principles explained in this walk-through may still be useful.

In this chapter, we will go step-by-step through the editing process, from opening the software to exporting your first completed film. By the end of the chapter and its associated video lesson, you should have enough knowledge to use Premiere Pro successfully to competently edit your projects. This walk-through cannot teach you everything about this very powerful and versatile software package, but it will explain the fundamentals upon which you can continue to build. Before beginning this walk-through, ensure that you have the following resources:

- At least two separate video clips, such as a multi-camera interview or different pieces of environmental footage. Still images, such as photographs or illustrations, can also be used. If you have been completing the tasks assigned during each of the video lessons, you should have collected ample material by this point.

- At least one audio file not already associated with a video clip. This can include music, sound effects, or audio captured separately from a video file (such as the audio recorded by a lavaliere microphone during an interview or a commentary track).

- A folder on your device that contains all the relevant audio-visual files. This is not essential, but storing all your material in one location will simplify the editing process.

Step One: How to Start a New Project

1. Open the Adobe Premiere Software.
2. Select 'New Project'.
3. In the new window that appears, you will be able to give your project a name. Set the Display Format to 'Timecode'. Set the Audio Format to 'Audio Samples'. Set the Capture Format to 'HDV'.
4. Click 'OK' to create your project.

23. *Editing Workflow in Adobe Premiere Pro* 235

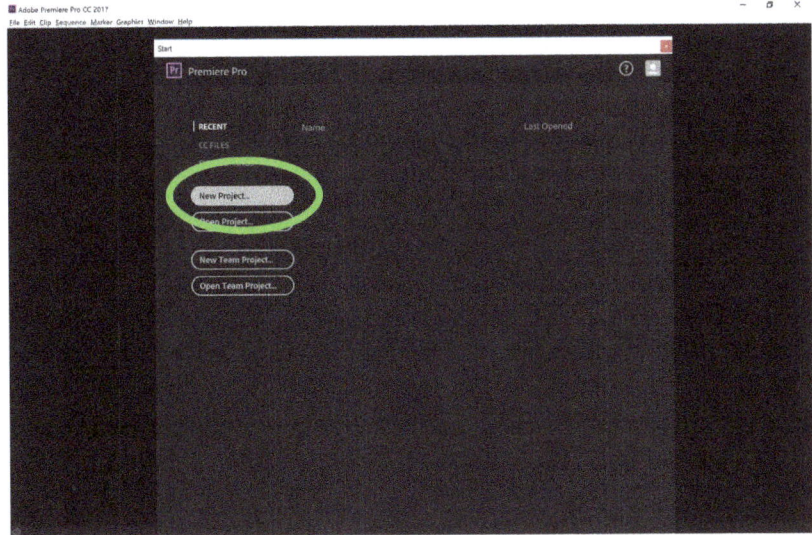

Fig. 61. Select "New Project" to begin.

Step Two: Get to Know the Premiere Workspace

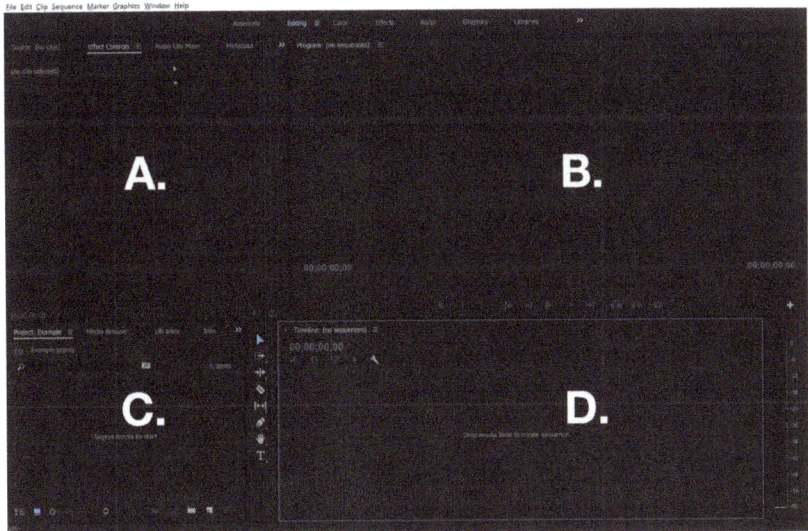

Fig. 62. The four main working areas in Premiere Pro.

1. The workspace in Premiere is generally divided into four sections:
   a. Effect Control Window.
   b. Preview Window.
   c. Media Bin.
   d. Timeline.
2. The Timeline is the space where audio and video clips are placed, manipulated, and edited together. The raw materials for your project (unedited audio and video) are stored in the Media Bin. The Preview Window will allow you to watch (preview) your project. The Effect Control Window allows you to manipulate aspects of your clips, such as their transparency level or position on the screen.

Step Three: Import Video and Audio Clips into your Project

3. Go to 'File'. Select 'Import'.
4. Navigate to the folder where you have stored your raw audio and visual files and select the file(s) you wish to import. Click 'OK'.
5. Your chosen file(s) will now appear in the Media Bin.
6. You are now ready to start editing your clips.

23. Editing Workflow in Adobe Premiere Pro

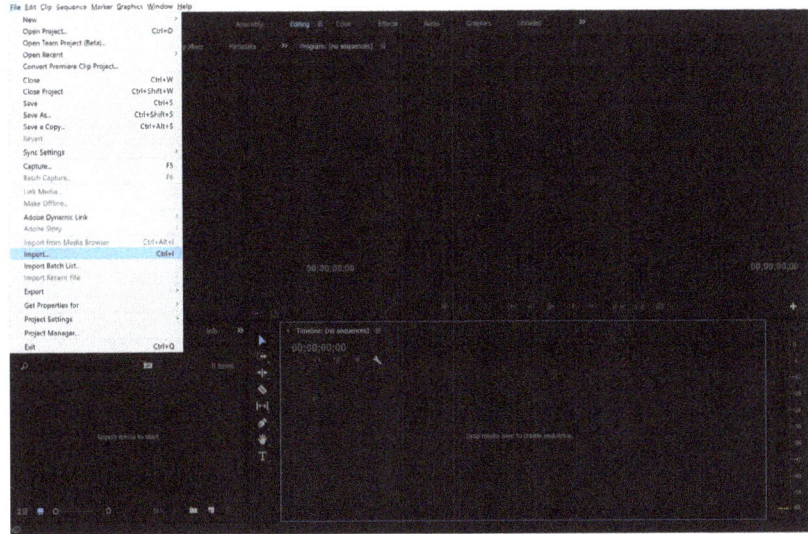

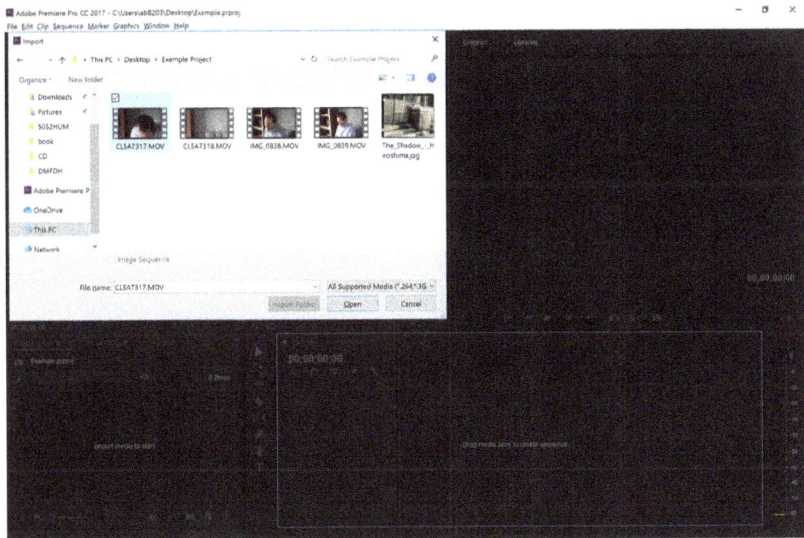

Figs. 63–64 Importing footage, audio and still images.

Step Four: Move Clips into your Timeline

Fig. 65. Moving footage from your project folder into your timeline.

1. Select the clip you wish to import into your film from the Media Bin.

2. Click the file and drag it into the Timeline window.

3. The clip will now appear in the Timeline window in its full, unedited form. Note that your Preview Window will now display a still image from the start of this video file.

4. At the top of the Timeline is a blue arrow. This arrow is connected to a long, thin blue line, which cuts vertically through your Timeline. It should be located at the time stamp 00:00.

    a. Click the blue arrow and drag it along your Timeline.

    b. See how the Preview Window changes as you begin scanning through your footage.

    c. Press the space bar. This will begin playback of whatever is in your Timeline. Note how the preview begins wherever the blue bar is located. Press the space bar again to stop the video from playing.

23. Editing Workflow in Adobe Premiere Pro

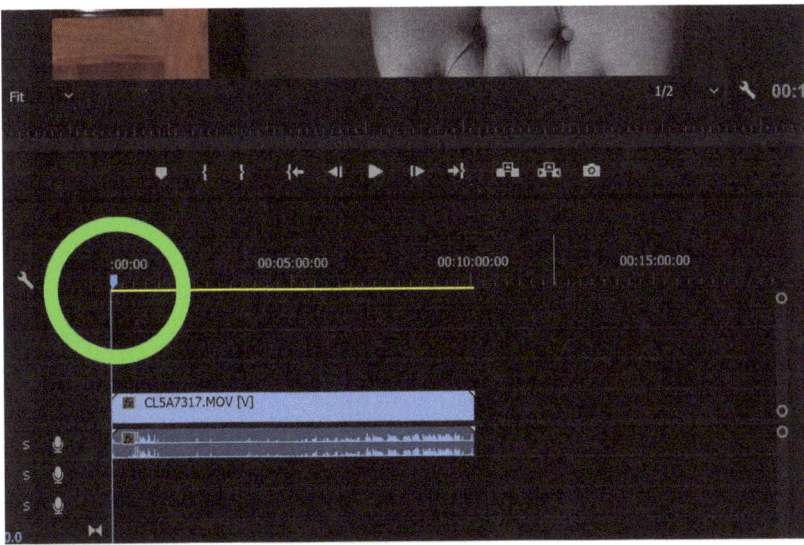

Figs. 66–67 Moving this blue bar will allow you to scroll through your project.

Step Five: Shorten a Clip

Fig. 68. The arrow cursor will allow you to easily select different parts of your project and begin manipulating them.

1. Select the arrow-shaped cursor from the toolbar.
2. Move your cursor to the end of the clip. Note how, as the cursor hovers over the end of the clip, it changes into a red bar with an arrow, which should face towards the left (see Figure 69).

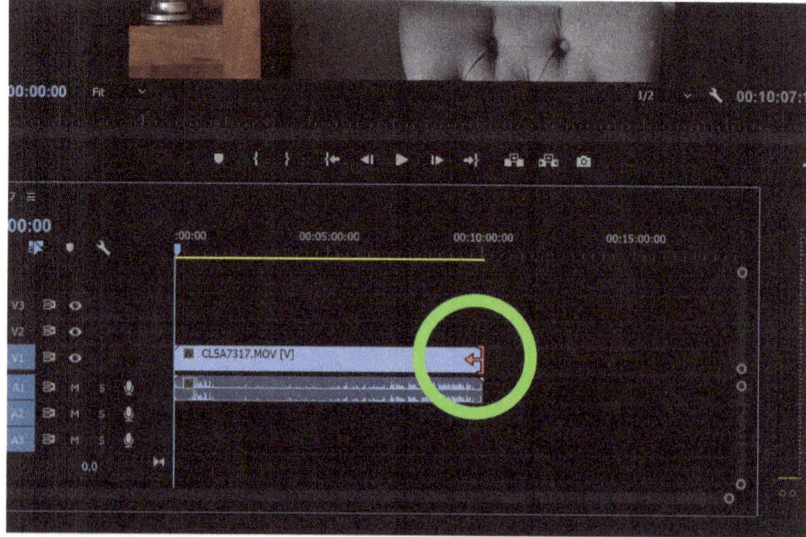

Fig. 69. Hovering the cursor over the end of a clip will allow you to shorten it.

23. *Editing Workflow in Adobe Premiere Pro*   241

3. Click the end of the clip. Drag your mouse to the left. Note how the clip begins to shrink as you drag your mouse.

   a. Move your cursor over the grey bar located at the bottom of your Timeline. At the end of this bar is a small, circular handle. Click this and drag it to the right or to the left. Note how the Timeline zooms in and out, allowing you to judge timing more accurately. Repeat step three until you shorten your clip to the desired length. Zoom in and out of the time as required to ensure that you have shortened it to the correct timestamp.

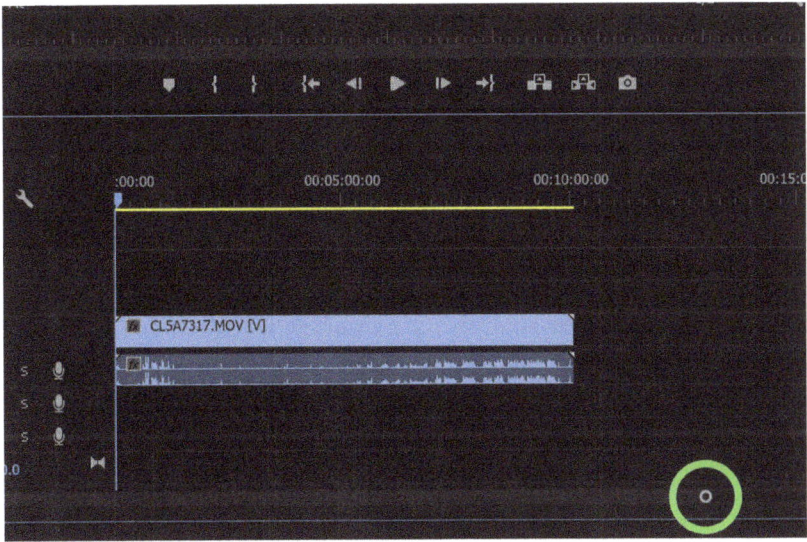

Fig. 70. Clicking and dragging this handle will allow you to zoom in and zoom out of your project.

Step Six: Moving Clips Around the Timeline

Fig. 71. Click on individual components within your timeline to rearrange them.

1. Select the clip you wish to move — click and hold with your left mouse button.

2. As you hold the left mouse button down, drag the clip left or right. Move it to the desired location in your Timeline.

Step Seven: Cutting Between Clips

1. Select a new clip from the Media Bin. Drag it into your Timeline (see step four). Drag the clip onto a different layer from that of the first clip. This will prevent you from accidentally overwriting clips you have already edited. Note how clips can exist concurrently in the timeline, if they are placed on separate layers. Clips are typically composed of a video and an audio element. The video element will appear in the top half of the Timeline. The audio element will appear in the bottom half of the Timeline. When importing a new clip to your Timeline, ensure that neither the audio nor the video elements overwrite any of your previously imported work.

23. Editing Workflow in Adobe Premiere Pro

You can change the layer on which a clip sits by clicking and dragging it up or down on the timeline.

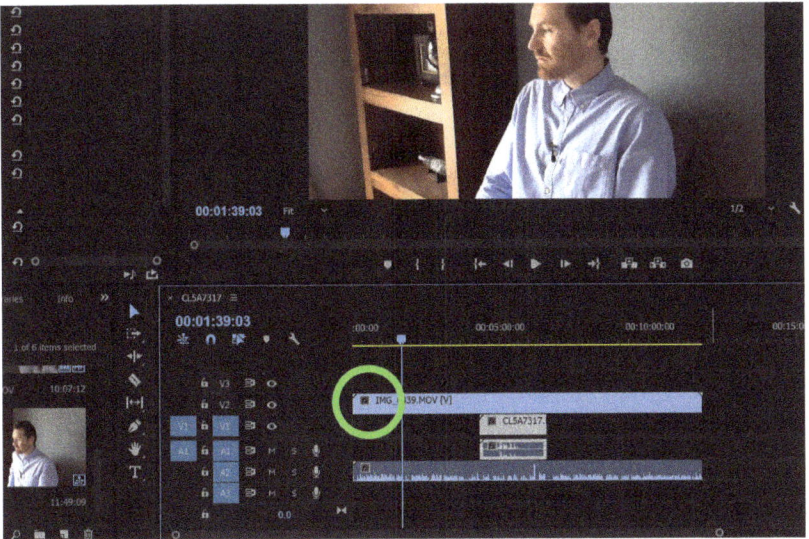

Fig. 72. Video and audio components can be stacked in the timeline and then rearranged accordingly.

2. Reduce the new clip to the desired length, as per the instructions in step five. You can shorten a clip by hovering your cursor over its start and/or its end point. This will allow you to remove unwanted material that appears from the first or the second half of the clip.

3. Once you have shortened the new clip to its desired length, click and drag it to the desired location on your Timeline.

4. Click and drag one of your clips so that its starting point lines up with the end of the other. You can now move your clip onto the same 'layer' as the first. This will make it easier to connect the two clips (see Figures 73 and 74).

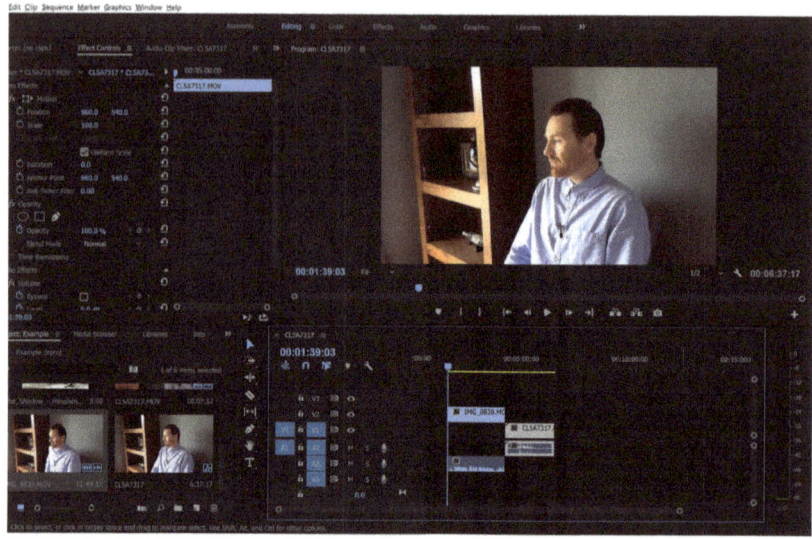

Fig. 73. Above: one clip will finish playing and the second will then immediately commence.

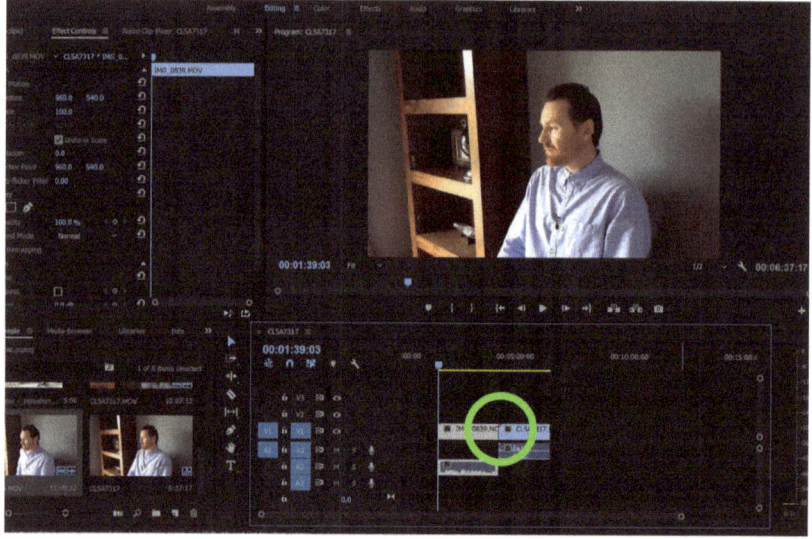

Fig. 74. By moving edited clips onto the same layer, you can keep your project well organised.

5. Select all your clips simultaneously (press the shift key and then select each clip with your mouse) and then drag them to

the start of your Timeline. This will ensure that your videos begin playing at the start of your project.

6. Drag the blue arrow to the start of your sequence and press the space bar to preview the sequence.

Step Eight: Remove Unwanted Sound-tracks

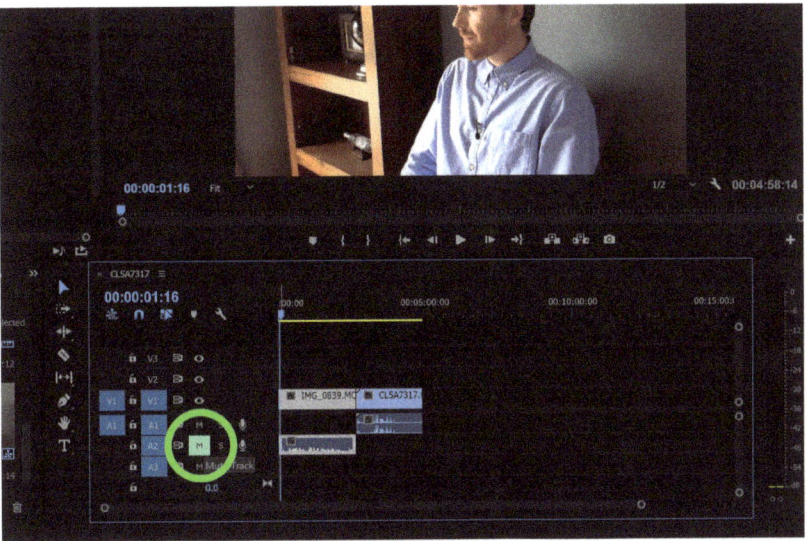

Fig. 75. The "M" button will mute all sounds on a given layer.

1. In order to remove any unwanted sound (such as the sound recorded by your camera's built-in microphone), locate the audio track on which the sound is located.

2. To the left of that track click the 'M' icon. This audio track will now be muted — all tracks which appear on this layer will be muted.

3. To delete the audio entirely, right-click the sound-track in question. This will bring up a pop-up menu. Within this menu, select 'Unlink'. This will allow you to edit the audio and video from the original clip independently.

4. Select the audio you wish to remove and press the 'Delete' key.

Step Nine: Add a New Sound-track

1. Import your audio files into your Media Bin (see step three).
2. Locate the audio file, click it, and drag it to the desired position in your Timeline (see step four).
3. Shorten the track in exactly the same way that you would a video file (see step five) and place it in the desired location within your Timeline (see step six).
4. Press the space bar to preview the result. Note that as your video plays, any (unmuted) audio files through which the blue bar passes will play simultaneously. This will allow you to layer sounds and create a custom soundscape.
5. All audio files will play at their default volumes. To balance the audio, right-click one of the files and select 'Audio Gain'.

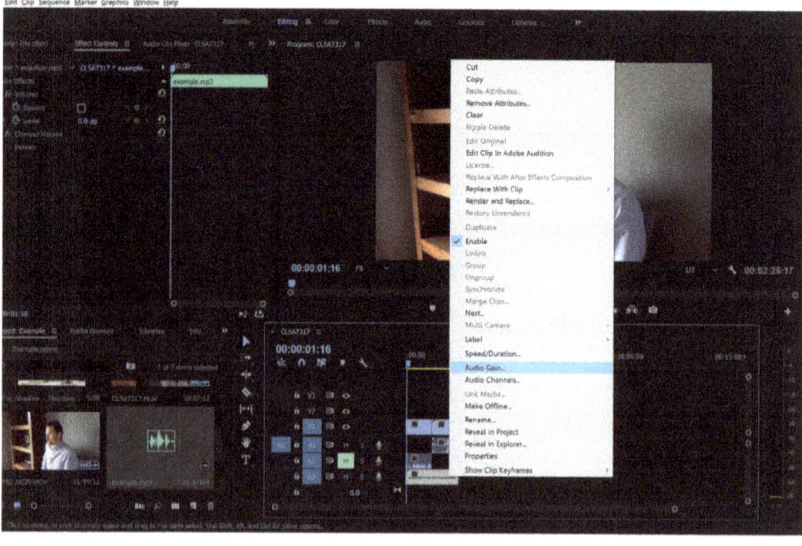

Fig. 76. Right click on a clip to bring up this menu. Selecting 'audio gain' will allow you to adjust its default volume.

6. To increase the volume of the selected clip, enter a positive value (above zero).
7. To decrease the volume of that clip, enter a negative value (below zero) into the 'Audio Gain' properties window.

8. Click 'OK' to apply these changes.
9. Press the spacebar to preview your project.
10. Repeat as necessary to gain the desired effect.
11. For a tutorial on syncing externally recorded audio with a video clip, please see the video lesson included in this chapter.

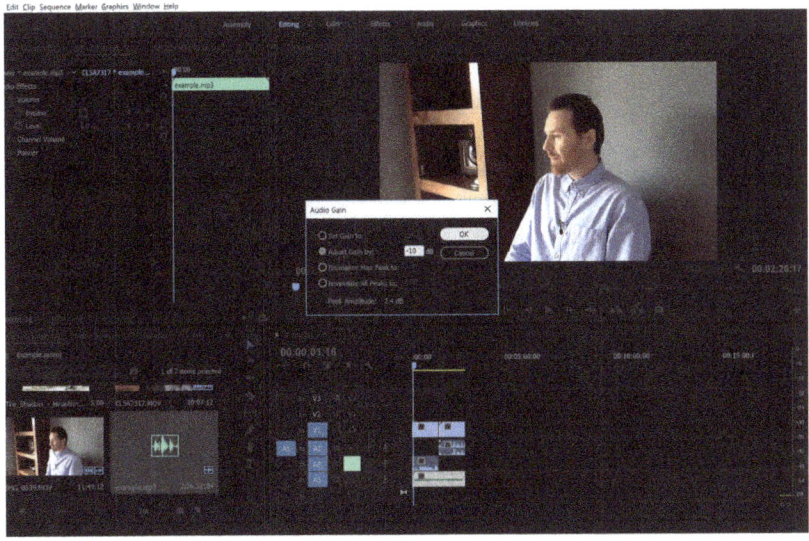

Fig. 77. Entering a negative value will reduce the default volume. Entering a positive value will increase it.

Step Ten: Add On-Screen Text

1. From the toolbar, select the 'Type Tool'.

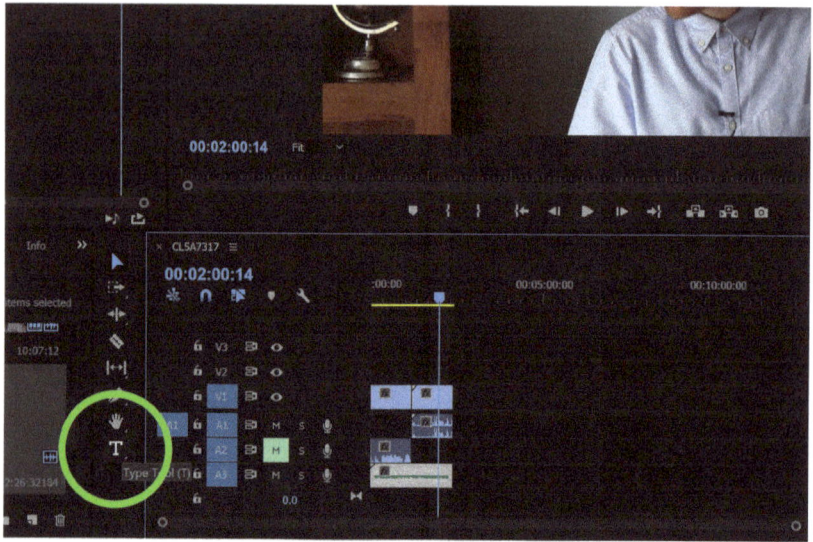

Fig. 78. Select the Text tool to generate on-screen captions.

2. Move your mouse cursor into the Preview Window. Left-click over the area where you wish your text to appear.

3. Left-click and enter the desired text. You can change the size, style, and font of the text by highlighting it and adjusting the desired properties located within the Effects Control Panel.

23. Editing Workflow in Adobe Premiere Pro

Fig. 79. Select "Effect Controls" to edit the text you have placed in a sequence.

4. In your Timeline, a new element will appear. This contains your text. Lengthen and shorten this element in order to control the duration for which the text will appear on screen. You can manipulate the element as you would any other visual element in your Timeline.

5. Note that text will only be overlaid onto a video clip if the text element is placed on a layer above the video. Premiere Pro stacks layers so that elements that appear on the uppermost layers appear above those that are stacked below it (see Figure 80).

Fig. 80. The text you have created will appear in the timeline as its own discreet entity. This can be manipulated in the same way as any other visual component in your timeline.

Step Eleven: Saving Your Project

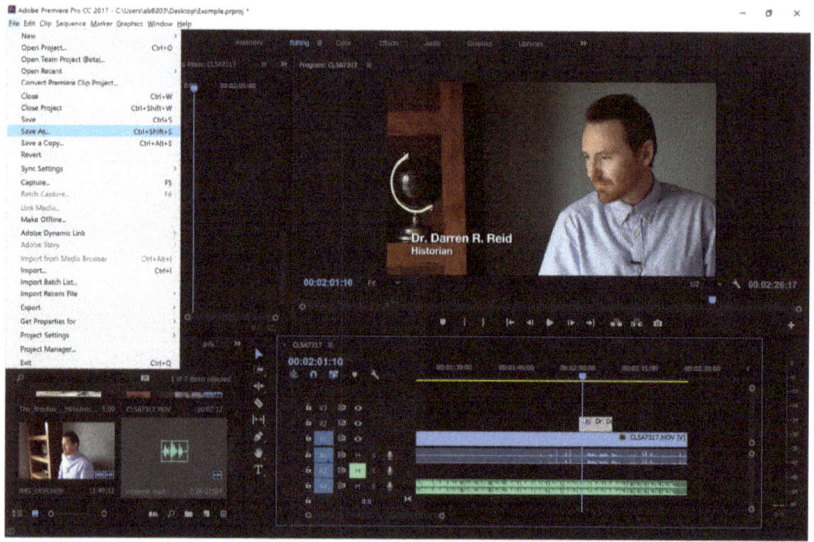

Fig. 81. Save your project regularly in order to avoid losing hours of work.

1. Go to 'File'. Select 'Save As'.
2. Select the location where you want to save your project.
3. Give your file a name.
4. Click 'OK'.
5. Note that the file you have created is a not a video file — the audio, video, and other elements have not yet been encoded. To create a video file that you can share you will need to 'export' your project.
6. To open a work-in-progress project, go to 'File'. Select 'Open'. Select the project file you wish to resume editing and click 'OK'.

Step Twelve: Exporting Your Project

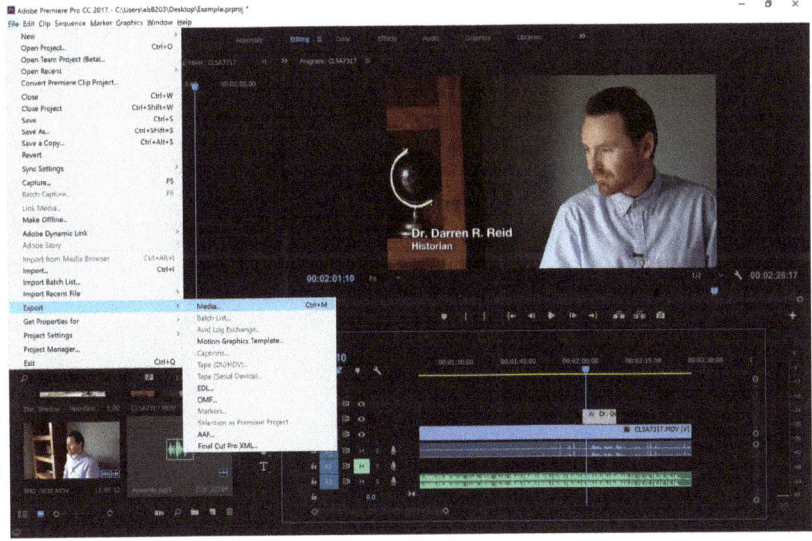

Fig. 82. Export your project to create a video file that you can share.

1. Go to 'File'. Select 'Export'. Select 'Media'.
2. A window that open. Select your preferred file format. H.264 is a widely used video standard that will produce a high-quality video with a reasonable file size.

252   *Documentary Making for Digital Humanists*

3. Click the blue text adjacent to 'Output Name' and select your desired file name and the location on your computer where you wish the encoded video to be stored.

4. Select the 'Video' tab. Scroll down within this window until you find the 'Bitrate' sliders.

5. A bit rate of 10–20Mbs will produce a high-quality video. You can lower the bitrate to reduce the file size. This may also reduce the quality of your exported file.

6. Click the 'Export' button. This will begin the process of encoding your project into a stand-alone video.

7. Please note that, depending upon the length and complexity of your project, the exporting process can take some time to complete.

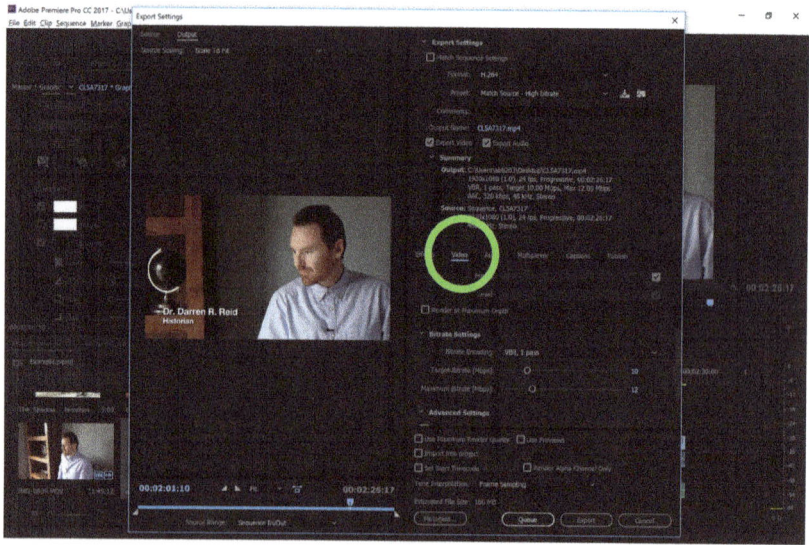

Fig. 83. Under the "Video" tab you will be able to define the settings for your exported file.

23. Editing Workflow in Adobe Premiere Pro

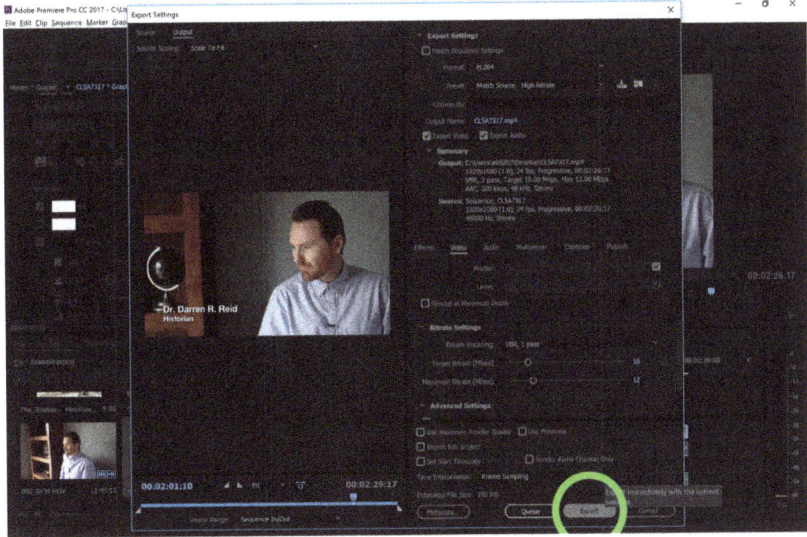

Fig. 84. Select "Export" to begin the process of turning your project into a completed video file.

Step Thirteen (Optional): Colour-Grading Your Project

Fig. 85 Watch this video lesson for an in-depth introduction to colour-grading in Adobe After Effects. http://hdl.handle.net/20.500.12434/2313fcf0

Colour-grading is the process of rebalancing or stylizing the look of your film. It is an in-depth process that requires significant practice. To accomplish this, please watch the above video lesson which will walk you through the fundamentals of the process using Adobe After Effects.

24. Distribution and Dissemination

Academic conventions for humanists remain rooted in the practices that matured in the nineteenth century. Academic histories are written, sometimes presented, but almost always disseminated via the written word, and even though quills have been replaced by typewriters, which were then replaced by word-processors and computers, the dominant dissemination practice of the historian has remained largely unchanged. Humanists write articles and books, disseminated by academic journals and publishers. Academic documentaries do not easily fit into this schema easily.

This raises some interesting questions for scholars who break from this convention and set out to produce academic films. The existing platforms of dissemination — books and academic journals — remain largely incompatible with the medium. The academic documentary is consumed on screens, but the question remains as to whose screens and where; in digital or physical spaces. Academic documentaries are currently obliged, at least at present, to find new ways to reach their target audience. This is both a challenge and an opportunity. A work in a new medium is necessarily disruptive and poses new methodological questions. Academic film also creates new opportunities to reach beyond the specialised readership of traditional academic texts.

In the absence of convention, you have the chance to propose and experiment with new conventions. How might one's work be peer-reviewed, or its impact measured? Clearly, as the producer of an academic piece, you must be recognised for your contribution.

When approaching the distribution process, you should consider the following questions:

© 2021 Darren R. Reid and Brett Sanders, CC BY-NC 4.0   https://doi.org/10.11647/OBP.0255.24

1. Who is the intended audience for this piece?
2. Where does that audience exist or congregate, in both online and offline spaces?
3. What will be required to speak directly to that audience?
4. What message would activate interest in your film among that audience?
5. Who are the gatekeepers who control or limit access to your desired audience? What message can spark the interest of these gatekeepers; why should they promote your project or help you to raise awareness?
6. Will your film work better in mobile-focused digital spaces (such as YouTube); in the home of the intended audience (via a digital streaming service); or in a curated event or exhibition (such as a screening)?

By answering these questions, you will be in a position to begin constructing a tailored dissemination strategy for your work. Such strategies will likely vary from the dominant dissemination strategies in your field. This is no bad thing and the opportunity to reach new audiences in new ways should be embraced.

Theatrical Release

By identifying an audience and the spaces where it exists and/or congregates, potential avenues for the film's release can likewise be identified quickly. For *Looking for Charlie*, a film about the history of the silent era, lovers of cinema were identified as a core audience. Online, these groups congregated in various internet forums and social-media groups. Offline, such individuals attended film festivals, the cinema, and cinema museums. Such venues created a clear path through which we could reach an audience most likely to respond to our work. Whilst not all academic documentaries require a theatrical presentation, *Looking for Charlie* is about the history of cinema, is a feature-length production, and has high production values. It was appropriate that it become an exhibition piece, shown in public spaces as part of a larger, immersive experience.

We wanted to exhibit the film in a series of physical spaces, to open up broader discussions about the themes and issues raised by our work as part of a larger series of events. As a documentary about the history of film, it made intellectual sense to attempt a limited theatrical run for *Looking for Charlie*; to have audiences engage with our work in the same way that they would engage with the works of Charlie Chaplin or Buster Keaton. A standard theatrical release was, of course, unlikely. Such endeavours require extensive planning, the cooperation of numerous theatres who perceive mass market appeal in the work, and, most importantly, a significant marketing budget to drive traffic into the cinemas in question. It is not enough merely to arrange a screening and hope that an audience will materialise. It is absolutely necessary to create awareness, crafting a message that is compelling enough to drive an audience to see your work.

Despite the difficulties associated with any type of theatrical release, we nonetheless set about creating an exhibition roadshow. The idea was simple: identify venues that would have some sort of natural synergy with our subject and begin building a series of screenings and events around those locations. In each location we would introduce our film and host a question-and-answer session. To drive our marketing narrative, we worked to produce a consistent body of artwork to promote the film, and a common tagline or message designed to accurately describe it to potential audience members: 'A film about the dark side of the silent era, from Charlie Chaplin and Buster Keaton to the forgotten clowns who inspired them'.

In order to reach a wider audience, a promotional campaign, which included local radio, television, posters, and flyers, was conducted. The flyer (see Figure 86) was produced using Photoshop and printed on high-quality paper — the quality of the design and the thickness and weight of the paper were important in reflecting the professional nature of the film's production. The same design was used on the posters; the consistency of the message and the symmetry of the promotion was of fundamental importance. In fact, extracting key parts of the film's message was key to gaining favourable press coverage. The main themes that played out across the promotion were:

- Appealing to people's nostalgia for the silent era.

- Offering a deeper understanding of the art: the DNA of comedy.
- Humanising performers.
- The mental health themes covered within the film.

Our premiere event occurred in the city of Coventry, which had recently been awarded the accolade of City of Culture 2021. As this is our home city, we were able to pay particular attention to this screening. We selected a high-quality, large-capacity venue, which we turned into a 'pop-up cinema and museum'. We took this approach for a number of reasons. Firstly, our choice of venue allowed us to sidestep the politics of the modern film industry, with which every dedicated cinema must contend. Rather than potentially seeing our film as a nuisance — something to be accommodated between more profitable Hollywood fare — our chosen venue embraced our project, making it one of their featured events. As such, they were incentivised to make the most of the experience, recognising that it would add to the fabric of what that venue already offered. We were able to build a larger event around the screening, allowing us to create a more fully realised, immersive experience. A pop-up museum was added, as was a screening of a Buster Keaton film with a live piano accompaniment, and the sale of cocktails from the era to complement the screening of our film.

We supported our premiere with extensive promotion, much of which took the form of high-quality posters and flyers which we distributed to local businesses. We particularly targeted those businesses and spaces that our target audience frequented. We also reached out to the press and were covered extensively by local newspapers, radio, and the BBC. Turning a bar into a pop-up museum was a novel idea, which generated a lot of attention — as did our film's focus on Charlie Chaplin, whose name and legacy continues to attract interest from a wide cross-section of people. Indeed, whilst our initial marketing focused upon college-educated people aged twenty-five to forty-five, the broad reach of the interviews we conducted with organisations such as the BBC demonstrated that college-educated over-fifties were another viable target audience.

The film's premiere was a resounding success. Many more people than we had anticipated attended the event, resulting in a packed venue.

24. *Distribution and Dissemination* 259

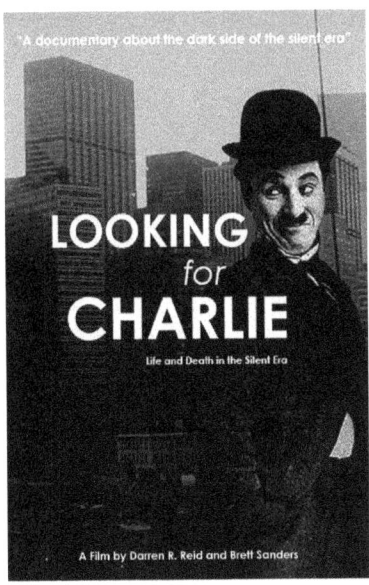

Fig. 86. Poster for *Looking for Charlie: Life and Death in the Silent Era*. This project was distributed as an 'event' film through a series of screenings presented by the filmmakers.

It also provided us with a model for how we could reach audiences in the future, as well as feedback on what aspects of our marketing message worked (and what did not). From here we continued to roll out the film, one screening at a time, picking venues that had a natural synergy with our subject, or those to which we could add entertainment and intellectual value. The result was a series of shows that allowed us to engage with a number of high-quality audiences with a deep interest in our subject and the main themes of our work.

The *Looking for Charlie* roadshow illuminated some core lessons about managing a film as an exhibition-style release. Significant promotional work is always required. Organising a screening is only one part of a much larger process, which involves creating awareness as well as the desire among potential audience members to attend a screening. On one occasion we were hosted by an organisation who had little interest in promoting our screening. It was a new organisation, which had yet to establish trust with its own customers, so organic footfall was light whilst targeted footfall (largely thanks to the dearth of promotion) was likewise sparse. Compared and contrasted with our other events, which were appropriately managed and promoted with a consistent, core message, the difference was striking.

We also learned that our core message had to be refined. Despite making a documentary about comedians, our film focused on depression and mental illness. It was, therefore, important that our potential audience understood what this film was (and what it was not). Word about our events had to be spread effectively in online *and* offline spaces. We had to construct a model of our imagined audience member: who were they; how old were they; what were their interests; what would make them want to attend our event? The subject of our film appealed to two distinct groups — older men and women (fifty years and older) who had a lifelong relationship with the subjects of our film (particularly Chaplin and Keaton). The other group was university-educated twenty-five to forty-five-year-olds who particularly enjoyed the consumption of retro-themed products and vintage culture.

To maximise the impact of our roadshow, we also produced a guestbook to which we invited audience members to contribute. We included some questions that we asked our audience to consider: 'In what ways did the film help you to learn more about the roots of Chaplin's comedy?' and 'In what ways did this film help you to reframe your knowledge of Vaudeville and the early silent era?'. Answers to these questions helped us to measure the impact and success of our film, whilst creating empirically based feedback for future academic work. This information, combined with the knowledge we gained from our roadshow, provided us with a wealth of knowledge that we could utilise as a part of our digital distribution model, ensuring that we can effectively target future potential audiences.

Digital Streaming

The growth of online streaming services such as Netflix, Amazon Prime, and YouTube has created new opportunities for scholars to reach very broad audiences. In reality, however, access to these channels is limited, and their broad reach may not make them appropriate for niche academic areas. Services such as Netflix tend to cultivate relationships with distributors who can offer them a catalogue of materials, rather than independent filmmakers who can typically offer them only a limited volume of content. Whilst this does not make it impossible for you to access these distribution channels, it does severely limit opportunities in this space.

In order to appear on the leading digital streaming platforms, you will need to find a distributor who has built, or who will attempt to build, a relationship with that platform. You will then have to sign over a significant portion of your film's rights. After all of this, your piece *might* appear on the desired streaming service. Alternatively, a distribution aggregator's services can be employed. Aggregators are a type of distributor who charge for their service. They collect a variety of related films into packages, which they then offer to online streaming services. If your film is part of a package picked up by a streaming service, it will appear in its catalogue. Again, there are no guarantees. Unlike a regular distribution deal, however, it is the filmmaker who must pay the aggregator (rather than the distributor paying the filmmaker) for the *possibility* of being picked up by a streaming service. In both of these cases, you are unlikely to be paid well for your work.

Fig. 87. *Keepers of the Forest* was released primarily through online streaming services. It has been screened in Brazil, where its subject matter is most relevant, but its primary international channels of dissemination are Amazon Prime and YouTube. https://youtu.be/ZywE92bDCrQ.

Gaining distribution through large-scale streaming services may prove an insurmountable challenge. In that case, a more viable option may be embracing free-to-access distribution spaces, which allow for long-term, organic audience accumulation. Services such as YouTube offer a range of distribution opportunities, which can be combined, if desired, with existing channels of digital scholarly publication. YouTube may not foster a particularly academic audience but the ability to embed content from the site into other online spaces provides a zero-cost method of integrating scholarly films into online journals and publications. Such works should, of course, speak to the intellectual aims, goals, and standards of the academic entity with which you wish to work. Scholarly presses are increasingly open to having discussions about the inclusion of audio/video content within their (digital) pages.

Whatever distribution space you choose, it is important to understand that publishing a piece does not mean that it will find an audience. Whether you release via a free-to-access platform such as YouTube or a premium streaming service such as Netflix, it is your responsibility to identify your audience, understand how your film will add value to them, and seek them out. Do not assume that your audience will discover your work amid the vast amount of content vying for their attention in the online space. Your documentary may appeal to a distinct and underserved niche, but if that audience does not know your work exists (and if they cannot easily access it) it will struggle to find traction.

To that end, revisit the questions listed at the outset of this chapter and utilise them as fully as you can in the digital space. In addition, you might also consider the following questions: to which online communities do my intended audience belong; how do they use social media; how can I introduce them to my work in a way that will encourage them to engage with it?

Freely Accessible Digital Streaming

YouTube offers a free, easy, and accessible method of hosting videos online. There are, however, some drawbacks to the platform. Despite offering options to host HD videos, the service compresses the files that are uploaded to it. This can lower the quality and introduce unwanted visual artefacts. More problematically, the service tailors the quality of its videos to reflect the speed of the viewer's internet connection. Whilst this has advantages for the end-user, it can result in them viewing a downgraded version of your film, plagued by a lower than intended resolution or inferior sound quality. Your film might load at a faster speed, but the viewing experience will, for many, be inferior.

Despite this, YouTube remains the standard through which video content is consumed, particularly on mobile devices. Social networks such as Facebook and Twitter include video streaming and sharing services, making them ideal for simple, highly shareable (viral) clips. Social networks, however, are not built around a centralised, searchable database of publicly available video content. YouTube fills this niche and, as a result, it attracts an audience that is actively hoping to discover and consume video-based content which appeals to their interests. By placing your content on a site like YouTube, you make it comparatively easy for users to discover, particularly if your work services a specific niche not widely catered to on the site. In such cases, viewership may be small, but it is also likely to be engaged and appreciative.

Despite its apparent ubiquity, YouTube is not the only free-to-access, online streaming service that can be used to host your films. Vimeo, in particular, offers an alternative, which, for a small monthly fee, allows users to host full, non-compressed HD content which will not be downgraded to accommodate slower internet connections. In practical terms, this means that filmmakers are able to control the quality of their documentaries, removing one of the principal problems faced by producers of high-fidelity content on YouTube. Vimeo's audience is significantly smaller than YouTube's, however, and, as a result, there is less scope for an uploaded video to organically develop a large audience. If a film has been produced primarily for distribution through scholarly channels, as part of an open access article, for instance, it may be more important to control its visual and audio quality than it is to foster a broad audience. In such instances, Vimeo, rather than YouTube, may offer you a more suitable hosting solution.

Scholarly films are unlikely to attract a broad audience beyond their intended niche, unless specific effort has been expended upon creating a highly accessible survey of a popular topic. Still, there is always the potential (if not necessarily the likelihood) that works made available on sites such as YouTube and Vimeo will build a large audience. Scholarly films may not be particularly well suited to viral sharing, but these platforms nonetheless provide filmmakers with the opportunity, particularly over the long term, to grow sizeable audiences. Whether sought-after or not, filmmakers should be aware that works hosted on such services are likely to be seen outside of the academy and, as a result, comment, discussion, and discourse may be the result. On a freely accessible public forum such as YouTube, however, user comments can be destructive as well as constructive, so thought should be given to developing a strategy for dealing with provocative, unfair, bigoted, or prejudicial comments which might be posted onto your film's page.

Scholars may choose to produce documentaries specifically in order to communicate ideas to broader audiences. Such scholars should, however, manage their expectations. Producing and releasing a film, no matter its intellectual worth, does not guarantee that an audience of any significant size will engage with it. Whilst sites such as YouTube and Vimeo offer easy access to an international audience, a vast array of competing content on a variety of topics means that, unless one's film has very broad appeal, it is unlikely to gain a massive following. Still, it is possible to use such freely accessible channels to speak to a much larger audience than those attracted by many academic journals or scholarly monographs. As with a theatrical or premium digital-streaming release, you should ask fundamental questions about the audience you wish to attract. Who is your intended audience; how do they use sites like YouTube; what type of content are the looking for; what core message from you will attract them to your film?

In a fast-changing online landscape, user behaviour should not be taken as a given. Whatever the size of the audience you hope to attract, it is the responsibility of the filmmaker to identify the most appropriate distribution channels for their work, and the best way to engage their desired audience with their content. YouTube and Vimeo are often consumed in short bursts on small mobile devices, but the rise of Smart TVs and devices such as Apple TV and Google Chromecast allow that same content to be viewed in a very different way: on the user's TV, in the comfort of their home, where they might demand longer, more involved content.

Filmmakers should assume that potential viewers will not discover their films unless their existence is highlighted. Leverage your social networks, particularly public-facing profiles on sites such as Twitter, to communicate with potential viewers about your work. Create and update a profile of your intended audience and continue to reach out to them in a way that adds value to their lives: informative or entertaining social media posts that may or may not be related to your film. Endeavour not to over-promote your work; instead, use your film as a vehicle to drive broader conversations about its content whilst gently highlighting its existence and where it can be viewed.

Whatever approach you adopt for the dissemination of your film, understand that the distribution landscape is a fast-changing space with new developments occurring frequently. Rather than offering specific guidance, which is likely to become outmoded before it can be actioned, this chapter has instead sought to draw your attention to several broad approaches to the dissemination of your work. You, and only you, should be the ultimate author of your work's distribution model.

To accomplish that, you will need to develop a clear sense about what you wish to achieve. You will then need to consider your preferred audience, understanding where that audience resides and how you can effectively reach them with your work. You might also consider the places that this audience congregates in the real world and develop a method of reaching them there. Do you wish to screen your work in front of an audience; to what extent do you wish to interact with your audience; how do you wish these interactions to occur; is your work part of a larger, curated experience or do you expect your audience to consume it as part of a larger diet of bite-sized audio-visual content? Beginning to answer these questions will allow you to begin to understand how current distribution models can be used to most effectively to disseminate your work.

Ensure that you place the audience's experience at the heart of your model. Whilst the minutiae of the distribution landscape changes frequently, your audience should be relatively constant. Understand who you are making your film for, in order to devise the best path to connect this audience to your work. Keep your intended audience at the centre of your vision for dissemination: this will guide you far more effectively than any temporary market trend.

Bibliography

Abrams, Lynn, *Oral History Theory* (Abingdon: Routledge, 2016), https://doi.org/10.4324/9780203849033

Alan, David Brown (ed.), *Virtue and Beauty: Leonardo's Ginevra de' Benci and Renaissance Portraits of Women* (Princeton and London: Princeton University Press, 2001).

Altman, Rick, *Silent Film Sound* (New York: Columbia University Press, 2004), https://doi.org/10.1017/s0021875806211319

Andersson, Barry, *The DSLR Filmmaker's Handbook: Real-World Production Techniques. Second Edition* (Indianapolis: John Wiley and Sons, 2015).

Arnheim, Rober, *Film as Art* (Berkley and London: University of California Press, 1957).

Austin, Thomas and Wilma de Jong (eds), *Rethinking Documentary: New Perspectives, New Practices* (Maidenhead: Open University Press, 2008).

Baranowski, Andreas M., 'Effect of Camera Angle on Perception Trust and Attractiveness', *Empirical Studies of the Arts* 31:1 (2017), 1–11, https://doi.org/10.1177/0276237417710762

Barthes, Roland, *Camera Lucida* (New York: Hill and Wang, 1981).

Belkhir, Jean Ait and Christiane Charlemaine 'Race, Gender, and Class Lessons from Hurricane Katrina', *Race, Class and Gender*, 14: 1/2 (2007), 120–52.

Bernard, Sheila Curran, *Documentary Storytelling: Creative Nonfiction on Screen* (New York and London: Focal Press, 2014), https://doi.org/10.4324/9780080962320

Bernard, Sheila Curran Bernard, *Documentary Storytelling: Creative Non-Fiction on Screen. Fourth Edition* (New York: Focal Press, 2016).

Berry, David M. (ed.), *Understanding Digital Humanities* (New York: Palgrave Macmillan, 2012), https://doi.org/10.1057/9780230371934_1

Billinge, Sam, *The Practical Guide to Documentary Editing: Techniques for TV and Film* (New York: Routledge, 2017), https://doi.org/10.4324/9781315233123

Bowen, Christopher J. and Roy Thompson, *The Grammar of the Shot* (London and New York: Focal Press, 2013), https://doi.org/10.4324/9780240526096

Boyd, Don, 'We are all Filmmakers Now — and the Smith Review Must Recognise That', *The Guardian*, 25 September 2011, https://www.theguardian.com/commentisfree/2011/sep/25/all-film-makers-smith-review

Brady, Jacqueline, 'Cultivating Critical Eyes: Teaching 9/11 Through Video and Cinema', *Cinema Journal* 42 (2004), https://doi.org/10.1353/cj.2004.0002

Braidotti, Rosi, *The Post-Human* (Cambridge: Polity Press, 2013), pp. 143–85.

Brown, Blain, *Cinematography: Theory and Practice — Image Making for Cinematographers and Directors* (New York: Routledge, 2016); Mercado, *The Filmmaker's Eye*; and Sijll, *Cinematic Storytelling*, https://doi.org/10.4324/9780080958958

Brown, Christopher Leslie Brown "Foreword" in Winthrop D. Jordan *White Over Black: American Attitudes Towards the Negro, 1550–1812. Second Edition* (2012; Chapel Hill: University of North Carolina Press, 1969), https://doi.org/10.5149/9780807838686_jordan

Brown, Stuart L., foreword to *The Heroes Journey: Joseph Campbell on his Life and Work* by Joseph Campbell (New York: New World Library, 2003).

Callaghan, Barry, *Film-making* (London: Thames and Hudson, 1973).

Campbell, Joseph, *The Hero with a Thousand Faces. Third Edition* (New York: Pantheon Books, 1949; reprint, Novato: New World Library, 2008).

Cheng, Eric, *Aerial Photography and Videography Using Drones* (Berkeley: Peachpit Press, 2006).

Christiansen, Mark, *Adobe After Effects CC: Visual Effects and Compositing Studio Techniques* (Adobe: New York, 2014).

Coppolla, Francis Ford, *Live Cinema and its Techniques* (New York and London: Liverlight, 2017).

Cossar, Harper, 'The Shape of New Media: Aspect Ratios, and Digitextuality', *Journal of Film and Video* 61:4 (2009), 3–16, https://doi.org/10.5040/9781501311680.ch-023

Cousins, Mark, *The Story of Film* (London: Pavilion, 2011), https://doi.org/10.1093/screen/46.2.281

Cox, Alex, 'Not in Our Name', *The Guardian*, 9 July 2005, https://www.theguardian.com/books/2005/jul/09/featuresreviews.guardianreview12

Craven, David, 'Style Wars: David Craven in Conversation with...', *Circa* 21 (1986), 12–14, https://doi.org/10.2307/25556947

Crowdus, Gary, 'The Editing of Lawrence of Arabia', *Cinéaste* 34 (2009), 48–53.

Dancyger, Ken, *The Technique of Film and Video Editing: History, Theory, Practice* (New York and London: Focal Press, 2011), https://doi.org/10.4324/9780080475202

Dejong, William, Eric Knudsen, and Jerry Rothwell, *Creative Documentary Theory and Practice* (London and New York: Routledge, 2021), https://doi.org/10.4324/9781315834115-9

Doherty, Thomas, 'Film and History, Foxes and Hedgehogs', *OAH Magazine of History* 16 (2002), 13–15: https://doi.org/10.1093/maghis/16.4.13

Ebert, Roger, 'The Birth of a Nation Movie Review (1915)' *RogerEbert.com*, 30 March 2013, http://www.rogerebert.com/reviews/great-movie-the-birth-of-a-nation-1915

Favlo, Joseph D., 'Nature and Art in Machiavelli's The Prince', *Italica* 66 (1989), 323–32, https://doi.org/10.2307/479044

Ferri, Anthony J., *Willing Suspension of Disbelief: Poetic Faith in Film* (Lanham: Lexington Books, 2007), https://doi.org/10.5040/9781474218627.ch-001

Field, Syd, *Screenplay: The Foundations of Screenwriting* (New York: Random House, 2005).

Figgis, Mike, *Digital Filmmaking. Revised Edition* (London: Faber & Faber, 2014), https://doi.org/10.5040/9780571343508

Fenton, Hugh, *Cinematograph: Learn from a Master*, YouTube, 27 April 2012, https://www.youtube.com/watch?v=KwtpJ3T8eK4&t=7s

Forsyth, Mark, *The Elements of Eloquence* (London: Icon Books, 2013).

Fried, Michael, 'Barthes' *Punctum*', *Critical Inquiry* 31 (2005), 539–74, https://doi.org/10.2307/3651445

Galloway, Dayna, Kenneth B. McAlpine, and Paul Harris, 'From Michael Moore to JFK Reloaded: Towards a Working Model of Interactive Documentary', *Journal of Media Practice* 8 (2007), 325–39, https://doi.org/10.1386/jmpr.8.3.325/1

Gardner, Eileenr and Ronald G. Musto, *The Digital Humanities* (Cambridge and New York: Cambridge University Press, 2015), https://doi.org/10.1017/cbo9781139003865

Gilbert, Felixt, 'The Humanist Concept of the Prince and the Prince of Machiavelli', *The Journal of Modern History* 11 (1939), 449–83, https://doi.org/10.1086/236395

Glebas, Francis, *Directing the Story: Professional Storytelling and Storyboarding Techniques for Live Action and Animation* (New York and London: Focal Press, 2009), https://doi.org/10.4324/9780080928098

Goldberg, David Theo, *The Afterlife of the Humanities* (Irvine: University of California Humanities Research Institute, 2014), https://humafterlife.uchri.org

Gouwens, Kenneth Gouwens, 'Perceiving the Past: Renaissance Humanism after the "Cognitive Turn"', *The American Historical Review* 103 (1998), 55–82, https://doi.org/10.1086/ahr/103.1.55

Grele, Ronald J., *Envelopes of Sound: The Art of Oral History. Second Edition* (1985; New York: Greenwood Publishing, 1991).

Gunn, Simon and Lucy Faire (eds), *Research Methods for History* (Edinburgh: Edinburgh University Press, 2011).

Grove, Elliot, *Raindance Producers' Lab: Low-to-No Budget Filmmaking. Second Edition* (Burlington: Focal Press, 2014), https://doi.org/10.4324/9780240522197

Hampe, Barry, *Making Documentary Films and Reality Videos* (New York: Henry Holt and Company, 1997), pp. 279–83.

Harmon, Dan 'Story Structure 101–106', Channel 101 Wiki, http://channel101.wikia.com/wiki/Story_Structure_101:_Super_Basic_Shit

Harrison, Brian 'Oral history and recent political history', *Oral History* 1 (1972), 30–48.

Higher Education Academy (n.a.), *Historical Insights Focus on Research: Oral History* (Coventry: Warwick University Press, 2010).

Hill, Andy, *Scoring the Screen: The Secret Language of Film Music* (Milwaukee: Hal Leonard Books, 2017).

Hirose, Yoriko, Alan Kennedy, and Benjamin W. Tatler, 'Perception and Memory Across Viewpoint Changes in Moving Images', *Journal of Vision* 10:4 (2010), 1–19, https://doi.org/10.1167/10.4.2

Huhndorf, Shari M., 'Nanook and his Contemporaries: Imagining Eskimo Culture, 1897–1922', *Critical Inquiry* 27 (2000), 122–48, https://doi.org/10.1086/449001

Hurkman, Alexis Van, *Color Correction Handbook* (New York: Peachpit Press, 2014).

Irving, David K. and Peter W. Rea, *Producing and Directing Short Film and Video. Fifth Edition* (Burlington: Focal Press, 2015), https://doi.org/10.4324/9780080468419

Izhaki, Roey *Mixing Audio: Concepts, Practices, and Tools* (Burlington: Focal Press, 2013), https://doi.org/10.4324/9780080556154

Jacobs, Ronald N., 'Civil Society and Crisis: Culture, Discourse, and the Rodney King Beating', *American Journal of Sociology* 101 (1996), 1238–72, https://doi.org/10.1086/230822

Jaksic, Ivan, 'Oral History in the Americas', *The Journal of American History* 92 (1992), 590–600, https://doi.org/10.2307/2080049

Jolliffe, Genevieve and Andrew Zinnes, *The Documentary Filmmakers Handbook* (New York: Continuum, 2006), ttps://doi.org/10.5040/9781501340475

Kahn, Victoria, '*Virtù* and the Example of Agathocles in Machiavelli's *Prince* Representations' 13 (1986), 63–83, https://doi.org/10.2307/2928494

Karcher, Barbara C., 'Nanook of the North', *Teaching Sociology* 17 (1989), 268–69, https://doi.org/10.2307/1317496

Katz, Steve, *Film Directing: Shot by Shot* (Michigan: Michael Wiese, 1991)

Keast, Greg, *Shot Psychology: The Filmmaker's Guide for Enhancing Emotion and Meaning* (Honolulu: Kahala Press, 2014).

Kilborn, Richard and John Izod, *An Introduction to Television Documentary* (Manchester: Manchester University Press, 1997).

Kit, Borys, '"The Dark Knight Rises" Faces Big Problem: Audiences Can't Understand Villain', *Hollywood Reporter*, 20 December 2011, https://www.hollywoodreporter.com/heat-vision/dark-knight-rises-christian-bale-batman-tom-hardy-bane-275489

Krages, Bert, *Photography: The Art of Composition* (New York: Allworth Press, 2005).

Kobbé, Gustav 'The Smile of the "Mona Lisa"', *The Lotus Magazine* 8 (1916), 67–74.

Landau, David, *Lighting for Cinematography* (New York: Bloomsbury, 2014), https://doi.org/10.5040/9781501376146

Lee, Kevin B., 'Video-Essay: The Essay Film — Some Thoughts of Discontent', *Sight and Sound*, 22 May 2017, http://www.bfi.org.uk/news-opinion/sight-sound-magazine/features/deep-focus/video-essay-essay-film-some-thoughts

Lipschultz, Jeremy Harris, *Social Media Communication: Concepts, Practices, Data, Law, and Ethics* (New York: Routledge, 2015), https://doi.org/10.4324/9780429202834

Mackendrick, Alexander, *On Filmmaking* (London: Faber & Faber, 2006).

Mamet, David, *On Directing* (New York: Penguin, 1992), pp. 1–7, 26–47.

Marcus, Alan, 'Reappraising Riefenstahl's *Triumph of the Will*', *Film Studies* 4 (2004), 75–86, https://doi.org/10.7227/fs.4.5

Martin, Adrian, *Mise En Scene and Film Style* (New York: Palgrave Macmillan, 2014), https://doi.org/10.3366/film.2015.0054

Marwick, Arthur, *The Sixties* (Oxford: Oxford University Press, 2011).

McKee, Robert, *Story: Substance, Structure, Style, and the Principles of Screenwriting* (New York: Harper Collins, 1997).

McNeil, Patrick and Steve Chapman, *Research Methods* (London: Routledge, 2005), https://doi.org/10.4324/9780203463000

Mercado, Gustavo, *The Filmmaker's Eye: Learning (and Breaking) the Rules of Cinematic Composition* (New York and London: Focal Press, 2010).

Murch, Walter, Murch, *In the Blink of an Eye. Second Edition* (Los Angeles: Simlan-James, 2001).

Nesbet, Anne, *Savage Junctures: Sergei Eisenstein and the Shape of Thinking* (London and New York: I.B. Taurus, 2003), https://doi.org/10.5040/9780755604128

Nichols, Bill, *Introduction to Documentary Film* (Bloomington and Indianapolis: Indian University Press, 2001), pp. 99–137.

Papazian, Elizabeth and Caroline Eades (eds), *The Essay Film: Dialogue, Politics, Utopia* (London and New York: Wildflower Press, 2016), https://doi.org/10.7312/papa17694

Perkins, V.F., *Film as Film: Understanding and Judging Movies* (London: Viking, 1972).

Perks, Robert and Alistair Thomson *The Oral History Reader* (London and New York: Routledge, 1998).

Pierce, Greg and Gus Van Sant, *Andy Warhol's The Chelsea Girls* (New York: Distributed Art Publishers, 2018).

Prucha, Francis Paul review of *Bury My Heart at Wounded Knee: An Indian History of the American West* by Dee Brown, in *The American Historical Review* 77:2 (1972), 589–90.

Quinn, James (ed.), *Adventures in the Lives of Others: Ethical Dilemmas in Factual Filmmaking* (New York: I.B. Taurus, 2015), https://doi.org/10.5040/9780755695201.

Rabiger, Michael, *Directing: Film Techniques and Aesthetics. Third Edition* (London and New York: Focal Press, 2003), https://doi.org/10.4324/9780080472539.

Rabiger, Michael, *Directing the Documentary* (Abingdon: Focal Press, 1987), https://doi.org/10.4324/9781315768335.

Reid, Darren R., 'Silent Film Killed the Clown: Recovering the Lost Life and Silent Film of Marceline Orbes, the Suicidal Clown of the New York Hippodrome', *The Appendix* 2:4 (2014), http://theappendix.net/issues/2014/10/silent-film-killed-the-clown.

Rhode, Eric, *A History of Cinema from Its Origins to 1970* (London: Penguin, 1972).

Ritchie, Donald A., *Doing Oral History* (Oxford: Oxford University Press, 2015)

Rocchio, Vincent F., *Cinema of Anxiety: A Psychoanalysis of Italian Neorealism* (Austin: University of Texas Press, 1999).

Rodriguez, Robert, *Rebel Without a Crew: Or How a 23-Year-Old Filmmaker with $7000 Became a Hollywood Player* (London: Penguin, 1996).

Rosenston, Robert A., *Visions of the Past: The Challenge of Film to our Idea of History* (Cambridge: Harvard University Press, 1998).

Rosenston, Robert A., *History on Film, Film on History* (London and New York: Routledge, 2012).

Saltzman, Steve, *Music Editing for Film and Television: The Art and the Process* (Burlington: Focal Press, 2015), https://doi.org/10.4324/9780203582787

Sarris, Andrew, *You Ain't Heard Nothing Yet: The American Talking Film, History and Memory, 1927–49* (Oxford: Oxford University Press, 1998).

Schenk, Sonja and Ben Long, *The Digital Filmmaking Handbook* (Los Angeles: Foreing Films Publishing, 2017).

Schreibman, Susan, Ray Siemens, and John Unsworth (eds), *A New Companion to the Digital Humanities* (Chichester: John Wiley & Sons, 2016), https://doi.org/10.1002/9781118680605

Schuursmam Rolf 'The Historian as Filmmaker I' and John Greenville 'The Historian and Filmmaker II' in Paul Smith (ed.), *The Historian and Film* (London and New York: Cambridge University Press, 1976).

Scoppettuolo, Dion and Paul Saccone, *The Definitive Guide to Da Vinci Resolve* (Blackmagic Design: Port Melbourne, 2018).

Sherman, Sharon R., 'Bombing, Breakin', and Getting Down: The Folk and Popular Culture of Hip-Hop', *Western Folklore* 43 (1984), 287–93, https://doi.org/10.2307/1500122

Sides, Hampton, Foreword to *Bury My Heart at Wounded Knee: An Indian History of the American West* (1972) by Dee Brown (New York: Henry Holt and Company, 2007).

Sijil, Jennifer Van, *Cinematic Storytelling: The 100 Most Powerful Film Conventions Every Filmmaker Must Know* (Michigan: Michael Wiese, 2005).

Sorlin, Pierre, *The Film in History: Restaging the Past* (Oxford: Blackwell, 1980).

Smith, John Thomas, *Remarks on Rural Scenery* (London: Nathaniel Smith, 1797).

Smith, Paul (ed.), *The Historian and Film* (London and New York: Cambridge University Press, 1976).

Tarlton, Charles D., 'The Symbolism of Redemption and the Exorcism of Fortune in Machiavelli's *The Prince*', *The Review of Politics* 30 (1968), 323–48.

Thomson, Alistair, 'Four Paradigm Transformations in Oral History', *The Oral History Review* 34 (2007), 49–70, https://doi.org/10.1525/ohr.2007.34.1.49

Thomson, Paul, *The Voice of the Past: Oral History* (Oxford: Oxford University Press, 1988).

Thompson, Paul, *The Voice of the Past: Oral History. Third Edition* (Oxford and New York: Oxford University Press, 2000).

Toplin, Robert Brent (ed.), *Ken Burns's The Civil War: Historians Respond* (New York and Oxford: Oxford University Press, 1996).

Yamato, Jen, 'The Science of High Frame Rates, Or: Why "The Hobbit" Looks Bad at 48FPS', *Movieline*, 14 December 2012, http://movieline.com/2012/12/14/hobbit-high-frame-rate-science-48-frames-per-second

Wolsky, Tom, *From iMovie to Final Cut Pro X: Making the Creative Leap* (Focal Press, New York, 2017), https://doi.org/10.4324/9781315456294

Yorke, John, *Into the Woods: How Stories Work and Why We Tell Them* (London: Penguin, 2013).

Aftermath: A Portrait of a Nation Divided. Digital Stream. Directed by Brett Sanders and Darren R. Reid. Coventry: Red Something Media, 2016.

Amélie. Directed by Jean-Pierre Jeunet. UGC: Neuilly-sur-Seine, 2001.

Annie Hall. Directed by Woody Allen. Los Angeles: United Artists, 1977.

Atomic: Living in Dread and Promise. Directed by Mark Cousins. London: BBC, 2015.

Battleship Potemkin. Digital Stream. Directed by Serge Eisenstein. Moscow: Goskino, 1925.

Blade Runner. Directed by Ridley Scott. Los Angeles: Warner Bros, 1982.

Capitalism: A Love Story. Directed by Michael Moore. Los Angeles: The Weinstein Company, 2009.

Confessions of a Superhero. Directed by Matthew Ogens. Toronto: Cinema Vault, 2007.

Culloden. Directed by Peter Watkins. London: BFI, 1964.

Empire of Dreams. Directed by Kevin Burns and Edith Becker. Los Angeles: 20th Century Fox, 2004.

Exit Through the Gift Shop. Directed by Banksy. London: Revolver Entertainment, 2010.

F is for Fake. Directed by Orson Welles. London: Eureka Entertainment, 1973.

Fahrenheit 9/11. Directed by Michael Moore. Santa Monica: Lionsgate, 2004.

Keepers of the Forest: A Tribe of the Rainforests of Brazil. Directed by Darren R. Reid. Coventry: Studio Académé, 2019.

Los Angeles Plays Itself. Directed by Thom Andersen. New York: The Cinema Guild, 2004.

King of Kong: A Fistful of Quarters. Directed by Seth Gordon. New York: Picturehouse, 2007.

Nanook of the North. Directed by Robert J. Flaherty. New York: Pathé Exchange, 1922.

Looking for Charlie: Life and Death in the Silent Era. Directed by Darren R. Reid and Brett Sanders. Coventry: Studio Académé, 2018.

Manhattan. Directed by Woody Allen. Los Angeles: United Artists, 1979.

My Scientology Movie. Digital Stream. Directed by John Dower. London: BBC Films, 2015.

Nanook of the North. Directed by Robert J. Flaherty. New York: The Criterion Collection, 1999.

Napoleon Dynamite. Directed by Jared Hess. Hollywood: Paramount Pictures, 2004.

Professor Green: Suicide and Me. Digital Stream. Directed by Adam Jessel. London: BBC, 2015.

Raiders of the Lost Ark. Directed by Steven Spielberg. Hollywood: Paramount Pictures, 1981.

Reel Injun. Directed by Neil Diamond. Montreal: National Film Board of Canada, 2009.

Requiem for the American Dream. Directed by Peter Hutchison, Kelly Nyks, and Jared P. Scott. El Segundo: Gravitas Ventures, 2015.

Rodney King Tape. Filmed by George Holliday. Camcorder footage. Los Angeles, 1991.

Saltzman, Steve, *Music Editing for Film and Television: The Art and the Process* (Burlington: Focal Press, 2015).

Roger and Me. Directed by Michael Moore. Burbank: Warner Bros., 1989.

Roger Waters: The Wall. Directed by Sean Evans and Roger Waters. Universal City: Universal Pictures, 2014.

Star Wars: Return of the Jedi. Directed by Richard Marquand. Los Angeles: 20th Century Fox, 1983.

Style Wars. Directed by Tony Silver. New York: Public Arts Films, 1983.

The Fog of War: Eleven Lessons from the Life of Robert S. McNamara. Directed by Errol Morris. Culver City: Sony Pictures Classic, 2003.

The Dark Knight. Directed by Christopher Nolan. Burbank: Warner Bros., 2008.

The Godfather. Directed by Francis Ford Coppola. Hollywood: Paramount Pictures, 1972.

The Shawshank Redemption. Directed by Frank Darabont. Culver City: Columbia Pictures, 1994.

The Story of Film: An Odyssey. Directed by Mark Cousins. Edinburgh: Hopscotch Films, 2011.

Wonders of the Solar System. London: BBC, 2010.

Wonders of the Universe. London: BBC, 2011.

Illustrations

Fig. 1	An open access, ten-part video series is included as a part of this text. To watch the first video lesson, readers of the online edition of this text should click on the link reported below. Readers of the print book can access the video by scanning the above QR code. Users can do this by opening the camera application on their phone and taking a photograph of the QR code. http://hdl.handle.net/20.500.12434/0322725a	5
Fig. 2	Watch Looking for Charlie by clicking on the link below or scanning the QR code. *Looking for Charlie: Life and Death in the Silent Era*. Digital Stream. Directed by Darren R. Reid and Brett Sanders. Coventry: Studio Académé, 2018. http://www.darrenreidhistory.co.uk/stream-looking-for-charlie/	23
Fig. 3	The location titles in *Looking for Charlie* (seen here) pay homage to the caption style utilised in Marvel's *Captain America: Civil War* (2016). *Looking for Charlie* (00:25:38–00:25:46).	31
Fig. 4	Walking through downtown Manhattan at night. This sequence in *Looking for Charlie* required three moving cameras to follow two moving subjects, both of which were wired for sound, whilst a boom mic operator recorded the city ambience. This was not an easy sequence to shoot, but the result was visually dynamic, taking advantage of the naturally high production values that New York offers. *Looking for Charlie* (0:30:58–0:32:37).	36
Fig. 5	Watch the second lesson in our documentary-making course. http://hdl.handle.net/20.500.12434/43f4c29c	41
Fig. 6	Our smuggler crew prepare to ascend the Seaton Cliffs in Arbroath.	49
Fig. 7	The scenery around the town of Arbroath is inherently dramatic, adding significant production value to any scene shot there. No tall ships were required to give this scene a sense of drama.	49

Fig. 8	Watch the trailer for *Looking for Charlie*. Scan the QR code or visit http://hdl.handle.net/20.500.12434/2313fcf2	56
Fig. 9	Shooting on location at Cirencester, behind the scenes at Gifford's Circus for *Looking for Charlie*. L-R, Darren R. Reid, Brett Sanders, and our subject for the day, Tweedy, a professional clown.	57
Fig. 10	*Nanook of the North* (1922), directed by Robert J. Flaherty.	67
Fig. 11	Watch the next lesson in our documentary-making course. https://hdl.handle.net/20.500.12434/c9b0ef48	75
Fig. 12	With only a small additional investment, you can transform the equipment you already own into a basic documentary-making kit. You can utilise your existing smartphone if it is able to capture HD or 4K footage. An older model can be paired with a lavaliere microphone and used as a sound recorder. An inexpensive smartphone adaptor would allow the phone to be connected to a tripod or to one of the stabilisation devices pictured (a gimbal and C-grip). Excluding the cost of the phone(s), the equipment in this setup could be purchased for a total of approximately $120. Pictured, from left to right, top to bottom: tripod, phone holder with tripod adaptor, mobile phone, lavaliere microphone, second mobile phone, gimbal, c-grip.	77
Fig. 13	Assembled over time, a DSLR kit's cost can be staggered. This setup was assembled over two years, and cost approximately $800. The camera is a Nikon D5500. It has 18–55mm, 55–200mm, and 50mm lenses alongside a range of filters, a lens hood, and wide-angle and macro adaptors. A gimbal allows for smooth handheld footage, as does a C-grip. A smartphone with a compatible lavaliere microphone helps to round out this kit. Pictured, from left to right, top to bottom: tripod, c-grip, directional microphone, LED light panel, LED filters, focus pull, lens, lavaliere microphone, a pair of lenses, cold shoes, Nikon D5500, lens, mobile phone grip, assorted lens filters, mobile phone.	78
Fig. 14	Watch the next lesson in the video series. http://hdl.handle.net/20.500.12434/1956f791	81
Fig. 15	The 'Rule of Thirds' grid is frequently used to shape filmic compositions.	84
Fig. 16	This photograph makes little use of the grid, its subject having been centred without regard for the ways in which the axes of the grid might add tension to the frame.	85

Fig. 17	By moving the subject off-centre and lining them up along one of the 1/3 axes, a degree of tension and imbalance is added to this composition. There is now space into which the subject can look and there is a clearer sense of compositional clarity. Even in a still photograph, the viewer is primed to expect the subject to move from left to right, through the vacant space within the frame.	85
Fig. 18	For interviews, try lining up one of your subject's eyes with one of the intersections of the upper axes, as seen in this image.	86
Fig. 19	Watch the next lesson in the video series. http://hdl.handle.net/20.500.12434/92a4bc2b	93
Fig. 20	Two subjects standing approximately eight feet apart, photographed using an 18mm lens. Note how small many of the background details are. All rights reserved.	96
Fig. 21	The same two subjects, standing in the same positions, photographed using a 50mm lens. Note how the background subject now appears much closer to the foreground subject. Note also how the background details have increased in size. All rights reserved.	96
Fig. 22	When photographed in 200mm, the background subject (upon whom the focus has now been pulled) appears very close to the foreground subject. Also note how close the environmental background details appear relative to our subjects. The space in this frame has been severely compressed. All rights reserved.	97
Fig. 23	Watch the video lesson on shot composition. http://hdl.handle.net/20.500.12434/18da6176	103
Fig. 24	The subject's head is pressed against the top of the frame, giving the shot an unsatisfying feel.	104
Fig. 25	An over-abundance of head room is similarly unsatisfying to the eye. All Rights Reserved.	105
Fig. 26	A small space between the top of the head and the top of the frame, however, feels appropriate.	106
Fig. 27	A lack of looking room makes a frame spatially unclear.	106
Fig. 28	Despite the subject not having moved position, the addition of looking room makes greater visual sense.	107
Fig. 29	When shooting an interview, cameras should be positioned on one side of the 'axis' only.	108
Fig. 30	Two cameras photographing the same object.	109
Fig. 31	The cameras should be at least 30° apart, or the audience may become aware of the cut between these different angles.	109

Fig. 32	The framing of this shot is of a notably poorer quality than the framing in the rest of the film.	111
Fig. 33	By zooming in on the footage and reframing the results, a more effective alternative composition reveals itself. This version of the shot was not included in the final cut of the film.	112
Fig. 34	In Frank Darabont's *The Shawshank Redemption*, the triumphant finale sees the camera pan back as it looks down on the protagonist, his arms outstretched. The edge of the frame frequently represents the limits of the observable cinematic universe to the viewer. We know that the subject in the above photograph exists in a space that extends far beyond the limits of this frame — but the edge of the frame, and the subject's relationship to it, nonetheless impacts how an audience respond to the shot. In Darabont's film the frame is not static, as it is in the above homage. The camera movement serves symbolically to free Andy in a way that cannot be replicated in still photography.	113
Fig. 35	*Aftermath: A Portrait of a Nation Divided*, directed by Brett Sanders and Darren R. Reid (0:31–0:38).	115
Fig. 36	*Aftermath: A Portrait of a Nation Divided*, directed by Brett Sanders and Darren R. Reid (3:51–4:06).	118
Fig. 37	*Aftermath: A Portrait of a Nation Divided*, directed by Brett Sanders and Darren R. Reid (3:51–4:06).	118
Fig. 38	The low-angle shot replicates the perspective of a child looking up at an adult, implying strength in the subject.	122
Fig. 39	The high-angle shot, which replicates the perspective of an adult looking down upon a child, implies vulnerability.	122
Fig. 40	From *Triumph of the Will* (1935), directed by Leni Riefenstahl (1:02:55–1:08:02).	122
Fig. 41	A close-up will allow your audience to read subtle facial expressions and micro gestures not otherwise evident in mid-shots (and certainly not in wide shots).	123
Fig. 42	The standard 16:9 aspect ratio will fill the entirety of a modern widescreen television.	124
Fig. 43	The 4:3 aspect ratio tends to evoke the era of early Hollywood. This aspect ratio is useful for generating a sense of nostalgia.	126
Fig. 44	A 21:9 aspect ratio is common in modern cinema. This aspect ratio is useful in evoking the sense of hyper-reality that so often accompanies modern films.	126
Fig. 45	Watch the video lesson on conducting interviews. http://hdl.handle.net/20.500.12434/47ac0bf7	127

Fig. 46a	The sound wave fits comfortably within the recordable field.	149
Fig. 46b	The device's recording sensitivity is too high, or the microphone is too close to a sound source.	149
Fig. 47	Backlit by the setting sun, the sky is perfectly clear and detailed whilst the subject is cast into shadow. To bring out the subject's features, a separate light source, aimed at them, would have been required.	159
Fig. 48	This LED panel cost less than $60 and can be mounted to a stand. It comes with a number of different filters, which can be used to defuse the light whilst increasing or decreasing the light's colour temperature.	161
Fig. 49	A homemade rig, assembled over time from inexpensive but effective component parts. A C-grip forms the basis of it. Cold-shoe extenders allow for external accessories, including lights and microphones, to be added to the rig. This is a handheld setup that has been attached to a tripod for stationary shots without needing to be disassembled.	167
Fig. 50	Tracking shot captured in New York by a camera operator following two subjects. *Looking for Charlie* (0:30:58–0:32:37).	168
Fig. 51	A folded tripod placed across the shoulder can serve as a crude shoulder stabiliser. When using such a setup, walk with bent knees, raising and lowering your feet so that they remain parallel to the ground. Do not push up using the ball of your foot to avoid ruining your shot with a bounce.	169
Figs. 52–53	The tripod dolly: the tripod's front legs remain stationary as the entire set up is pushed forward. The tripod's head is loosened so that the camera can remain perpendicular to the ground.	173
Fig. 54	Watch the video lesson on conducting interviews. https://hdl.handle.net/20.500.12434/c9b0163c	175
Fig. 55	Watch *Aftermath: A Portrait of a Nation Divided*. https://youtu.be/bU1wf4UIt-o.	189
Fig. 56	The three acts of a production each has a distinctive role to play. The first act sets out the premise, core ideas, and principle argument (or line of inquiry) for the piece. The second act engages in the substantive investigation and analysis. The third act brings those core ideas and arguments to their fundamental conclusion.	204
Fig. 57	The documentary embryo overlaid onto the three act structure.	216
Fig. 58	The Odessa Steps sequence. *Battleship Potemkin* (1925). Directed by Sergei Eisenstein (0:48:15–0:56:03).	222

Fig. 59	A still from one of the earliest films. The difference between the highlights (light areas) and shadows (dark areas) captured by celluloid are stark and evident here. This effect can be emulated by deepening shadows and blowing out highlights in post-production software. *Train Pulling into a Station* (1895), directed by Auguste and Louis Lumière.	228
Fig. 60	Watch this video lesson for an in-depth introduction to editing in Adobe Premiere Pro. https://hdl.handle.net/20.500.12434/6ff71a81	233
Fig. 61	Select "New Project" to begin.	235
Fig. 62	The four main working areas in Premiere Pro.	235
Figs. 63–64	Importing footage, audio and still images.	237
Fig. 65	Moving footage from your project folder into your timeline.	238
Figs. 66–67	Moving this blue bar will allow you to scroll through your project.	239
Fig. 68	The arrow cursor will allow you to easily select different parts of your project and begin manipulating them.	240
Fig. 69	Hovering the cursor over the end of a clip will allow you to shorten it.	240
Fig. 70	Clicking and dragging this handle will allow you to zoom in and zoom out of your project.	241
Fig. 71	Click on individual components within your timeline to rearrange them.	242
Fig. 72	Video and audio components can be stacked in the timeline and then rearranged accordingly.	243
Fig. 73	Above: one clip will finish playing and the second will then immediately commence.	244
Fig. 74	By moving edited clips onto the same layer, you can keep your project well organised.	244
Fig. 75	The "M" button will mute all sounds on a given layer.	245
Fig. 76	Right click on a clip to bring up this menu. Selecting 'audio gain' will allow you to adjust its default volume.	246
Fig. 77	Entering a negative value will reduce the default volume. Entering a positive value will increase it.	247
Fig. 78	Select the Text tool to generate on-screen captions.	248
Fig. 79	Select "Effect Controls" to edit the text you have placed in a sequence.	249
Fig. 80	The text you have created will appear in the timeline as its own discreet entity. This can be manipulated in the same way as any other visual component in your timeline.	250

Fig. 81	Save your project regularly in order to avoid losing hours of work.	250
Fig. 82	Export your project to create a video file that you can share.	251
Fig. 83	Under the "Video" tab you will be able to define the settings for your exported file.	252
Fig. 84	Select "Export" to begin the process of turning your project into a completed video file.	253
Fig. 85	Watch this video lesson for an in-depth introduction to colour-grading in Adobe After Effects. http://hdl.handle.net/20.500.12434/2313fcf0	253
Fig. 86	Poster for *Looking for Charlie: Life and Death in the Silent Era*. This project was distributed as an 'event' film through a series of screenings presented by the filmmakers.	259
Fig. 87	*Keepers of the Forest* was released primarily through online streaming services. It has been screened in Brazil, where its subject matter is most relevant, but its primary international channels of dissemination are Amazon Prime and YouTube. https://youtu.be/ZywE92bDCrQ.	261

About the Team

Alessandra Tosi was the managing editor for this book.

Melissa Purkiss performed the copy-editing and proofreading.

Anna Gatti designed the cover. The cover was produced in InDesign using the Fontin font.

Luca Baffa typeset the book in InDesign and produced the paperback and hardback editions. The text font is Tex Gyre Pagella; the heading font is Californian FB. Luca produced the EPUB, MOBI, PDF, HTML, and XML editions — the conversion is performed with open source software freely available on our GitHub page (https://github.com/OpenBookPublishers).

This book need not end here...

Share

All our books — including the one you have just read — are free to access online so that students, researchers and members of the public who can't afford a printed edition will have access to the same ideas. This title will be accessed online by hundreds of readers each month across the globe: why not share the link so that someone you know is one of them?

This book and additional content is available at:

https://doi.org/10.11647/OBP.0255

Customise

Personalise your copy of this book or design new books using OBP and third-party material. Take chapters or whole books from our published list and make a special edition, a new anthology or an illuminating coursepack. Each customised edition will be produced as a paperback and a downloadable PDF.

Find out more at:

https://www.openbookpublishers.com/section/59/1

Like Open Book Publishers

Follow @OpenBookPublish

Read more at the Open Book Publishers **BLOG**

You may also be interested in:

Remote Capture
Digitising Documentary Heritage in Challenging Locations
Jody Butterworth, Andrew Pearson, Patrick Sutherland and Adam Farquhar (eds)

https://doi.org/10.11647/OBP.0138

Digital Humanities Pedagogy
Practices, Principles and Politics
Brett D. Hirsch (ed.)

https://doi.org/10.11647/OBP.0024

Photography in the Third Reich
Art, Physiognomy and Propaganda
Christopher Webster (ed.)

https://doi.org/10.11647/OBP.0202

www.ingramcontent.com/pod-product-compliance
Lightning Source LLC
Chambersburg PA
CBHW041731300426
44115CB00022B/2978